ENDANGERED VERTEBRATES

GARLAND REFERENCE LIBRARY
OF SOCIAL SCIENCE
(VOL. 480)

ENDANGERED VERTEBRATES

A Selected, Annotated Bibliography, 1981–1988

Sylva Baker

GARLAND PUBLISHING, INC. • NEW YORK & LONDON
1990

Library of Congress Cataloging-in-Publication Data

Baker, Sylva, 1932–
Endangered vertebrates: a selected, annotated bibliography, 1981–1988 / Sylva Baker.
p. cm. — (Garland reference library of social science; v. 480)
ISBN 0–8240–4796–6 (alk. paper)
1. Vertebrates—Bibliography. 2. Endangered species—Bibliography. I. Title. II. Series.
Z7996.V4B35 1990
[QL605]
016.596'0529—dc20 90–2863
CIP

Printed on acid-free, 250-year-life paper
Manufactured in the United States of America

TABLE OF CONTENTS

ACKNOWLEDGEMENTS

The compilation of this bibliography took me to many libraries where hard-working and dedicated staff would bring to me the many periodicals and books I was requesting. The interest they took in this project has made this work a communal effort. Most importantly, every librarian made me feel that this was a reference resource that they needed and so spurred me on to great efforts at compiling, selecting, and annotating.

The academic libraries of Swarthmore College, West Chester University, Hahnemann University, Cornell University and the University of Pennsylvania were fine general sources. The museum libraries of The Academy of Natural Sciences of Philadelphia and the American Museum of Natural History were the best sources for journals and monographic series.

Information networking through OCLC, interlibrary loan, and delivery services have clearly demonstrated that even my tiny, local public library, the Helen Kate Furness Library of Wallingford, Pennsylvania, can be an access point to national bibliographic resources.

Many librarians aided this project but the number precludes naming them all. But special thanks are owed to Janet Evans of the Pennsylvania Horticultural Society (we began this project together) for her constant support. Also Carol Spawn, Marcia Gross, Joan Bari, Dorothy Viola, Linda Rossi, and in particular, Frank Mattiacci, all from my former home base, The Academy of Natural Sciences of Philadelphia, were outstanding in their help. Frank Mattiacci, the Interlibrary Loan Assistant, kept the OCLC sub-system humming with my requests. Nina Root of the American Museum of Natural History and her staff were skillful advisors for use of their collection on-site and through interlibrary loan. Bernadette Callery of the New York Botanical Garden supplied some articles from journals not found elsewhere. Judy Baker of the Hahnemann University Library and Mrs. Otte of the Marple (PA) Public Library were considerate and forthcoming. Michael Durkin, Director of the Swarthmore College Library, can be proud of his student assistants at the Cornell Science Library.

The patience and tact of Phyllis Korper and Paula Ladenburg, Garland Publishing editors, are what kept this project going and I do thank them for it.

I have often read the appreciative statements of authors to their families but I never realized their significance until now. My father taught me to love reading. My mother urged me to finish this project. Without the support and encouragement of my husband, David, this volume would not be in your hands.

INTRODUCTION

"Endangered species" has become a code term that is understood widely to mean the decline in the abundance of a particular plant or animal species, a decline so significant that the survival of that organism is in doubt. Since Rachel Carson sounded the alarm about DDT in her book, SILENT SPRING, there has been a heightened public awareness of the loss of animal species. The intent of this bibliography is twofold: (1) to provide the student, researcher, and interested reader with a comprehensive selection of publications that appeared between 1981 and 1988 containing information identifying vertebrates whose existence as a species is endangered, threatened, vulnerable, rare, or recently extinct, and (2) to cite references that contain information about the natural history of these organisms. By "natural history" I mean the biology, behavior, ecology, classification, reproduction, geographic range, and diet of the animal.

DEFINITION OF *ENDANGERED, THREATENED, VULNERABLE, RARE AND EXTINCT*

The standards for classifying an animal as endangered, threatened, vulnerable, rare or even extinct vary among nations and international bodies.

Although various associations and governmental bodies have their own wording to define the terms "endangered, threatened, vulnerable, rare, and extinct" the phrasing would generally agree with the following:

Endangered A species (or taxa) whose numbers are so decreased throughout all or a significant portion of their range that they are likely to become extinct very soon unless the conditions causing their decline are changed.

Threatened (U.S.) or **Vulnerable** (IUCN) A species (or taxa) whose numbers are in the process of decreasing at a rate which makes it likely that they will move into the endangered category very shortly unless the causal conditions are not changed.

Rare Species with small populations that are no longer generally seen in the wild. They may be in a stable status. They are not "endangered" or "threatened" but may become so. Their range may be a limited area.

Extinct Species have not been seen in the wild during the past fifty years. CITES and IUCN use this definition.

KEEPERS OF THE LISTS

Private groups and governmental agencies keep lists of species they consider endangered or threatened. Organizations concerned with a single species, e.g., seals, or a whole class, e.g., birds, have their own definitions, collect data on population number, and issue lists. Nations keep lists. The United States and many of the states keep lists. The Convention for International Trade in Endangered Species (CITES), the result of an international treaty, maintains a list of animals, skins, and parts of animal bodies restricted from international trade and export by member nations.

The official list of species considered endangered or threatened by the U.S. government is maintained by the Fish and Wildlife Service (FWS), an agency within the Department of the Interior. Some states accept the FWS list, others maintain their own.

Two international organizations that play leading roles in obtaining recognition of "endangered" status are the IUCN – International Union for the Conservation of Nature and Natural Resources and the ICBP – International Council for Bird Preservation. The two groups work together and produce the very important *Red Data Books*.

As a result of the many definitions of the terms endangered or threatened, and to provide the broadest view, the scope of entries accepts the authors' definition of what they consider "endangered, threatened, vulnerable, rare or extinct."

In reviewing the literature that was published from 1981 to 1988 to be included in the bibliography, it became apparent rather quickly that the number of published articles and books on the entire subject is overwhelming. To include them all would result in an unwieldy (and expensive) book as the topic of "endangered species" touches many related concerns other than natural history or biology.

Interest in endangered species is part of the concern for the total environment. Recent climatic changes, e.g., hot summers, drought, reports of loss of the ozone layer in the Antarctic, and strong hurricanes, have re-stirred interest in pollution of the atmosphere, the water supply and land. The tragedy of oil spills such as that caused by the tanker *Valdez* in Alaska brings TV images of dead birds and seals into our homes. The decline in the number of species and the number of animals have been an "early warning signal" that something is wrong. There is much research underway to understand the causes of the decline in species abundance and methods of preventing extinctions. This activity has resulted in a seemingly limitless amount and variety of literature concerned with "endangered species" written from the perspective of habitat destruction, wildlife management, population restoration, laws and regulations regarding trade in endangered animals, the heated controversy over the value of captive breeding programs in zoos and sanctuaries versus wildlife preservation strategy, laboratory research in genetics, theoretical works on the need for biological diversity, and research to determine the "minimal viable population." All directly involve issues relating to "endangered species."

The references selected from the plethora of publications for this volume have tried to remain true to the stated goals: identification of endangered vertebrates and information about their natural history. However, some citations will include elements of the tangential issues described above in addition to the goals of selection. I have included articles and books dealing with preservation and habitat destruction, etc. only insofar as they deal also with identification and natural history.

ORGANIZATIONS AS INFORMATION SOURCES

The interest in "endangered species" and the environment is found in many groups and on many levels of expertise: advocacy groups, scientists, students, government agencies and international bodies. The literature that documents their activities is a cross section of the publishing world. Scientists are publishing reports in scientific journals and books. Technical reports sponsored by organizations and the government are distributed. Advocacy groups promote their cause by educating the public through magazines, pamphlets, newsletters, news releases and special reports that are useful to scientists and the interested public. Advocacy groups and scientific organizations also sponsor symposia and frequently publish the proceedings. Graduate dissertations are available.

ABSTRACTING AND INDEXING SERVICES

The review of the literature also revealed that the sources were distributed in many types of publications, most of which are covered by abstracting and indexing services. The A & I services (both on-line and in hard copy) are definitely useful but there is no single index that contains all the relevant references for this topic. *Biological Abstracts, Zoological Record, General Science Index, Readers' Guide, Magazine Index, National Technical Information Service (NTIS), Science Citation Index, Environmental Bibliography, Books in Print, Dissertation Abstracts* and GPO (Government Publications Office)publications are good indices for the scientific and popular literature, but there is a need for a single, easy-to-use bibliography to pull together bibliographic listings over a period of years and from a variety of sources that refer the reader to publications about specific species. To meet this need, the present volume focuses on popular and scientific references that identify vertebrates as endangered, vulnerable, threatened, rare, or recently extinct, and that describe their natural history, biology, ecology and population abundance.

CRITERIA FOR INCLUDING CITATION

Publications may be included if their primary concern is habitat, conservation policy, laws and legislation, wildlife management, reproduction efforts or extinction theories if they also contain significant information about individual vertebrate species, perhaps as case studies. Excluded are writings about paleontology extinctions unless their discussion bears on the issues of current or anticipated species extinction.

FORMAT OF MATERIAL INCLUDED

Criteria for the format of publications have been selected to yield for the adult reader substantial information. References are annotated if the titles are not self-explanatory or the additional information will enable the reader to judge the usefulness of the original documents, some of which may not be easy to locate. Unannotated titles I was unable to see will be so indicated. I included some books or articles without seeing them because the author or the publisher was known for contributions in the subject. A few publications were difficult to find as they are collected only by major research libraries and may require a costly and time-consuming interlibrary loan process. The reader is advised of the possibility of this difficulty.

Selections include books, technical reports, monographs, symposia, journal and magazine articles, government documents, and dissertations. Excluded are newsletters, one-page articles (usually), pamphlets, newspaper articles, and similar ephemeral material. (I would **very strongly** encourage every reader to use the *New York Times Index*. The newspaper is the most comprehensive source of current information for the general public.) Articles in magazines that are digests or weekly reviews of news are omitted because the original information has more than likely come from a lengthy article elsewhere cited. Unsigned articles are not included nor are "letters to the editor" or book reviews.

The geographic scope of the contents is world-wide but only those publications written in English are included with an occasional exception for publications that contain an abstract in English and refer to an animal not written about elsewhere.

In summary, the criteria for inclusion in this bibliography are: 1) the publication appeared between 1981 and 1988, 2) an endangered vertebrate is the primary or secondary subject, 3) authors or editors are named, 4) articles are at least 2 pages in length (although there will be exceptions in the case of animals for which there is little other information), 5) the whole text or an abstract is written in English, and 6) the writing is for adults.

TIPS FOR READERS

Articles and books which contain bibliographies are the preferred choice and the reader is advised to use these sources. Due to the abundance of potential listings for this volume and the need to limit the number to be included, usually the most recent publication by an author about the same species is included.

If the most recent article or book by an author contains a bibliography or list of references, publications listed in that bibliography may not be included herein.

If the author has published several articles about the same subject in different journals, the latest and most complete article will be cited. However, an author may have written more than one article about the same subject for publication but intend them for readers who differ in their level of interest or knowledge. I have included both articles if they appeared in the scientific and popular publications to enhance the value of this volume to more readers. Furthermore, both articles will be included if one is in a more widely distributed and easily accessible journal while the other may have only regional distribution and be more difficult to find. It is expected that the bibliography or notes of the latest article will refer to previous writings of the author. Therefore, earlier writings will not be cited here.

The reader is encouraged to use the bibliographies of the books and articles listed in this volume to gather an additional list of references.

In looking up references for "endangered species" in library catalogs and abstracting and indexing series, the reader is advised to look under "rare animals" or "rare birds." The use of "endangered species" as a subject heading is used in some indices but is not universal. (Note: reptiles and amphibians are often grouped together as herpetofauna.)

In cases where a "species" has several substantial articles or books included in the present volume, additional short references will not be included if they do not contribute additional data.

Science Citation Index would be a helpful source to identify the author's previous publications and those writings cited most often by others.

ORGANIZATIONS

The reader should be aware of publications of advocacy groups such as the Audubon Society, World Wildlife Fund, Greenpeace, the Sierra Club, Save the Whales, Nature Conservancy and World Birdwatch among many. Other organizations that carry out research monitoring studies at an authoritative level are the International Union for the Conservation of Nature and Natural Resources (IUCN) and the International Council for Bird Preservation (ICBP). These and similar organizations are listed in the Appendix. Their addresses and the names of publications they issue are also listed.

LIST OF ABBREVIATIONS

The use of acronyms and alphabetic descriptors to identify organizations and agencies active in the field is sufficiently prevalent to require a listing to guide the reader.

INDEX OF SCIENTIFIC NAMES

Scientific names of vertebrates identified (with a major exception – see below) in the publications are given. The key number of the record in which they are cited is printed after the name.

The exception to the index are those publications that are checklists or mention more than ten animals. Those books or articles are not indexed due to the overwhelming number of animals. This exception also applies to the **INDEX OF COMMON NAMES** (see below.)

INDEX OF COMMON NAMES

Common names of vertebrates identified in the listed publications (with the exception noted above) are given. The key number of the record in which they are cited is printed after the name.

GEOGRAPHIC LOCATION INDEX

Publications that identify a geographic location will be indexed with the number of the record appearing after the nation, state, or geographic area. Publications that are world-wide in scope are not indexed.

ABBREVIATIONS

AAZPA	American Association of Zoos, Parks, Aquaria
AOU	American Ornithologists' Union
ASC	Association of Systematic Collections
BBOC	Bulletin of the British Ornithologists' Club
BES	British Ecological Society
CITES	Convention on International Trade in Endangered Species
EDF	Environmental Defense Fund
ESL	U.S. Endangered Species List
FOTE	Friends of the Earth
FWS	United States Fish and Wildlife Service
FWS/OBS	Fish and Wildlife Service/Office of Biological Services
GPO	{U.S.} Government Printing Office
ICBP	International Council for Bird Preservation
ICF	International Crane Foundation
IFCB	International Foundation for Conservation of Birds
IIED	International Institute for Environment and Development
IUCN	International Union for Conservation of Nature and Natural Resources
IUCN/SSC	International Union for Conservation of Nature and Natural Resources/Species Survival Commission
IWC	International Whaling Commission
IWRB	International Waterfowl Research Bureau
NGS	National Geographic Society
NOAA	National Oceanic and Atmospheric Administration
NPCA	National Parks and Conservation Association
NRDC	Natural Resources Defense Council
NTIS	{U.S.} National Technical Information Service
NWF	National Wildlife Fund
RAOU	Royal Australasian Ornithological Union
SC	Sierra Club
SSC	Species Survival Commission (See IUCN)
WBSJ	World Bird Society of Japan
WCI	Wildlife Conservation International
WWF	World Wildlife Fund

Endangered Vertebrates

GENERAL SOURCES

1. Aiken, S.R., and C.H. Leigh. 1985. On the declining fauna of peninsular Malaysia in the post-colonial period. *Ambio* 14(1): 15-23.

   Depicts the process by which this biologically diverse region is facing the loss of natural habitats.

2. Allan, D. 1985. Threatened plants and animals of the world. *Environmental Conservation* 12(1): 79-81.

3. Anadu, P.A. 1987. Progress in the conservation of Nigeria's wildlife. *Biological Conservation* 41(4): 237-251.

   Lists extinct and endangered species in addition to national parks and game reserves.

4. Anadu, P.A., and J.F. Oates. 1982. *The status of wildlife in Bendal State, Nigeria with recommendations for its conservation.* A mimeographed report prepared for the Nigerian Government, the New York Zoological Society, and the World Wildlife Fund.

   Conservation policy. A very limited printing and circulation. May be difficult to find.

5. Anderson, D., and R. Grove, eds. 1988. *Conservation in Africa: people, policies and practice.* New York: Cambridge Univ. Press.

   Much more about the human interaction with species that are endangered and the social and political implications than biology but useful.

6. Andrewartha, H.G., and L.C. Birch. 1984. *The ecological web: more on the distribution and abundance of animals.* Chicago: Univ. of Chicago. Pr.

7. Association of Systematic Collections. 1983. *Federally controlled species.* Vol. II of *Controlled wildlife: a three-volume guide to U.S. wildlife laws and permit procedures.* Lawrence, Kansas: ASC.

   Lists of species include, for each, status under Endangered Species Act, Convention on International Trade in Endangered Species (CITES), and other laws and regulations.

8. Batelle Memorial Institute. 1985. *Threatened and endangered wildlife survey: Vacherie Dome area, Louisiana.* Report ONWI-543, January 1985. Columbus, OH: Office of Nuclear Waste Isolation, Batelle Memorial Institute; Springfield, VA: National Technical Information Service. (NTIS)

An example of a specialized report prepared for a government agency as a source of information about endangered species. Reports done under government contract are available through NTIS, the National Technical Information Service.

9. Bauer, P. 1983. Endangered Species (New York State). *The Conservationist* 37(Jan-Feb): 19-20.

   Useful to identify New York endangered list.

10. Bendiner, R. 1981. *The fall of the wild, rise of the zoo*. New York: E.P. Dutton.

    Breeding endangered species in zoos for their survival. Provides introduction to this practice and describes well known cases in a popular style.

11. Berger, T.C., A.M. Neuner, and S.R. Edwards. 1981. *Directory of federally controlled species*. Lawrence, Kansas: Association of Systematics Collections.

    This authoritative listing is an update of the initial 1979 publication issued by the ASC, the major American organization for scientific collections concerned with study of taxonomy.

    See also item #7.

12. Boyd, J.M. 1984. Nature conservation. *Proceedings of the Royal Society of Edinburgh,* Section B-Biological Sciences. 84: 295-335.

    Primarily a historical review of conservation efforts in Scotland. Notes species given habitat reserves for their survival, particularly seabirds.

13. Branson, B.A., D.F. Harker, Jr., J.M. Baskin, M.E. Medley, D.L. Batch, M.L. Warren, Jr., W.H. Davis, W.C. Houtcooper, and B. Monroe, Jr. 1981. Endangered, threatened, and rare animals and plants of Kentucky. *Transactions of the Kentucky Academy of Science*. 42(3-4): 77-89.

    Not seen.

14. Byrne, G. 1988. Strengthened endangered species act passes. *Science* 242(4876)Oct 14: 190.

    A one-page article containing useful information about the legislation that had lapsed in 1985.

15. Cadieux, C. 1981. *These are the endangered.* Drawings by Bob Hines. Washington, D.C.: Stone Wall Press; Brattleboro, Vt.: Dist. by Stephen Green.

    Excellent overall introduction to issue of endangered species. Describes the most well known endangered, provides the U.S. Endangered Species List, the CITES list and a directory of names and addresses of related organizations.

16. Chandler, W.J., and L. Labate, eds. 1988. *Audubon wildlife report 1988/ 1989*. San Diego, CA: Academic Press.

Another in this annual series. A feature of each year's edition is a look at a government agency responsible for wildlife management. This year's report examines the National Marine Fisheries.

17. Coffin, B., and L. Pfannmuller, eds. 1988. *Minnesota's endangered flora and fauna*. Illustrations by J.A. Janssens. Minneapolis: Univ. of Minnesota Press for the Natural Heritage and Nongame Wildlife Programs of the Division of Fish and Wildlife, Minnesota Department of Natural Resources.

Large book listing over 75 species that are either extinct, extirpated, endangered, threatened, or of special concern. Species' descriptions include preferred habitat, aid to identification, status and reason for status, and references.

18. Commission of the European Communities, Convention on International Trade in Endangered Species of Wild Fauna and Flora (CITES). 1984. *European Communities Annual Report to CITES*. Paris: Commission of the European Communities.

Consists of a cumulative report of all trade in skins or animal parts of specimens between the European community and the rest of the world that had been proscribed by CITES. Report conveys the difficulties of getting complete records in order to implement the convention.

19. Convention on International Trade in Endangered Species (CITES). 1981. *Identification manual*. Ed. P. Dollinger. Compiled with the advice and guidance of the Identification Manual Committee. Gland, Switz. 5 v. (loose-leaf) Vol. 1. Mammalia – v. 2. Aves – v. 3. Reptilia. Amphibia. Pisces. – v.4. Parts and derivatives I – v. 5. Parts and derivatives II. Gland, Switzerland: Secretariat of the Convention.

Manual for identification of endangered species as agreed to by convention. Each species receives a descriptive page that includes the following: the common and scientific name, a drawing, appearance, and geographic information.

20. Cortese, T.J., and B.A. Grosnek, comps. 1987. *Selected publication series of the U.S. Fish and Wildlife Service, 1889-1985*. Fish and Wildlife Service, Resource Publication 159. Washington, D.C.: U.S. Department of the Interior, Fish and Wildlife Service.

An annotated bibliography of 19 Research and Development Series' publications. The Fish and Wildlife Service has, over the years, published many series, issued under new and merged titles. This bibliography provides a useful list of series' titles in addition to making the various series understandable to the reader.

21. Coto, Z., and E.A. Sumardja, eds. 1984. *Wildlife ecology in Southeast Asia. Proceedings of the Symposium on Wildlife Ecology in Southeast Asia, June*

7-9, 1983. BIOTROP Special Publication No.21. Bogor, Indonesia.

Publication of the Southeast Asian Regional Center for Tropical Biology. Useful for a region whose fauna and flora are subject to development and population pressures.

22. Cromartie, W.J., ed. 1982. *New Jersey's endangered and threatened plants and animals*. Pomona, N.J.: Stockton State College, Center for Environmental Research.

    Papers from the Second Symposium on Endangered and Threatened Plants and Animals of New Jersey, February 28-March 1, 1981.

23. Cypher, E.A., R.H. Yahner, G.I. Storm, and B.L. Cypher. 1986. Flora and fauna survey in a proposed recreational area of Valley Forge National Historical Park. *Proceedings of the Pennsylvania Academy of Sciences* 60(1): 47-50.

    Survey includes identification of endangered species in proposed area.

24. Day, D. *The doomsday book of animals: a natural history of vanished species*. With a foreword by David Attenborough. 1981. New York: Viking.

    Extinct vertebrates. A large general reference work, colorfully illustrated. Popular and scientific names are included.

25. Diamond, J.M. 1984. "Normal" extinctions of isolated populations. In *Extinctions,* ed., M.H. Nitecki, 193-246. Chicago: Univ. of Chicago Press.

    A paper delivered at the Sixth Annual Spring Systematics Symposium held at the Field Museum of Natural History, 1983, and included in the publication. A scholarly investigation of extinctions that are the normal pattern of populations on islands and archipelagos with case studies of birds and mammals.

26. DiSilvestro, R.L., and A.S. Eno., eds. 1986. *Audubon wildlife report 1986*. New York: National Audubon Society. (Distr. by Academic Press)

    General guide to activities, organizations, publications and government agencies concerned with wildlife. Accounts of selected species are quite informative. Good reference work. Appendix has directory to all government agencies relating to wildlife programs.

27. DiSilvestro, R.L., ed. 1987. *Audubon Wildlife Report 1987*. New York: National Audubon Society (Dist. by Academic Press.)

28. Dunlap, T.R. 1988. *Saving America's wildlife*. Princeton: Princeton Univ. Press.

    History of preservation efforts since 1880 but concentrating on the most recent period. Some natural history of the wolf and the red wolf is provided during a case history report of a program to restore them to the wild. Primarily it is a history of the attitudes and politics of wild animal management.

29. Ehrlich, P.R., and A. Ehrlich. 1981. *Extinction: The causes and consequences of the disappearance of species.* New York: Random House, Ballantine, 1983 (pap.)

Strong argument for preservation of non-human species in the description of the linkage of habitats and all forms of life (ecosystems). Case histories of well-known examples of extinctions and endangerment are given.

30. Faaborg, J.R., and W.J. Arendt. 1985. *Wildlife assessments in the Caribbean.* Rio Piedras, Puerto Rico: U.S. Forest Service, Southern Forest Experiment Station, Institute of Tropical Forestry.

Essentially a book of data concerning the study of habitat requirements of each animal species, and those species sensitive to habitat alteration. Authors identify species of particular importance and suggest species requiring further investigation.

31. Federal Wildlife Permit Office. 1983. *Annual report for 1982: Convention on International Trade in Endangered Species of Wild Fauna and Flora.*

Distributed by F&WS as FWS/WPO-83/02. Distributed by NTIS. PB84-146158. A technical report that reveals the operations of CITES.

32. Fjeldsa, J. 1983. Vertebrates of the Junin area in central Peru. *Steenstrupta* 8(14): 285-298.

Unseen but included because the geographic area is rarely "covered" in other publications.

33. Franklin, K.E. 1987. Endangered species: where to from here? *American Forests* 93 (11-12) Nov-Dec: 57-60, 74-76.

Author says the Peregrine falcon, red-cockaded woodpecker, grizzly bear, and the Puerto Rican parrot would have become extinct if it were not for the Endangered Species Act. Fine overview of the mechanics of the legislation, the need for greater funding, and the role of the U.S. Forest Service. Current funding permits 50 species to be added to the list yearly but there are 3800 candidates awaiting listing.

34. Fritts, T.H. et al. 1983. *Turtles, birds and mammals in the northern Gulf of Mexico and nearby Atlantic waters; an overview based on aerial surveys of OCS* [outer continental shelf] *areas with emphasis on oil and gas effects.* Washington, D.C.: U.S. Fish and Wildlife Service, Minerals Management Service. FWS/0BS-82/65.

Unseen. Noted in search of NTIS database.

35. Genoways, H., and F.J. Brenner, eds. 1985. *Species of special concern in Pennsylvania.* Compiled by the Pennsylvania Biological Survey. Special Publication of the Carnegie Museum of Natural History, No. 11. Pittsburgh: Carnegie Museum of Natural History.

A chapter devoted to each class of species. The volume's title says it all.

36. Gould League of Victoria (Australia). 1982. *The edge of extinction: Australian wildlife at risk*. Victoria: Gould League of Victoria.

In magazine format a generous description of species (arranged by habitat) is followed by an appendix that lists endangered species and their population size. A directory of conservation organizations in Australia is included.

37. Groves, R.H., and W.D.L. Ride, eds. 1982. *Species at risk: research in Australia*. Proceedings of a Symposium on the Biology of Rare and Endangered Species in Australia. Sponsored by the Australian Academy of Science and held in Canberra, November 25 and 26, 1981. Canberra: Australian Academy of Science and New York: Springer Verlag.

38. Hall, R.J. 1988. *An annotated bibliography of field studies*. Biological Report 88-2. Washington, D.C.: Fish and Wildlife Service.

Deals with a variety of species, many of whom are endangered, etc.

39. Holyroyd, G.L. et al. eds. 1987. Endangered species in the prairie provinces of Edmonton, Alberta, Canada. *Natural History Occasional Paper*, no. 9. Edmonton, Alberta: Historical Resources Division.

Unseen. Not easily available beyond the region.

40. Horigan, D.P. 1983. An update on endangered species in Hawaii from the Federal Register. *Elepaio* 44(3): 25-29.

41. Houtcooper, W.C., D.J. Ode, J.A. Pearson, and G.M. Vandel III. 1985. Rare animals and plants of South Dakota. *Prairie Nature* 17(3): 143-165.

Unseen but South Dakota not described elsewhere.

42. Howard, R., A. Perry, P. Herring, and J. Scott (Study Manager). 1984. *Endangered and threatened species on U.S. Air Force installations*. Fish and Wildlife Service/Office of Biological Services (FWS/OBS) 84/110. Washington, D.C.: U.S. Fish and Wildlife Service.

Publication consists of tables listing the vertebrates on U.S. Air Force bases that are classified as endangered, threatened, rare, protected and of special concern both on Federal and State lists. Scientific and common names are given.

43. International Convention Advisory Commission. 1981. *Convention on international trade in endangered species of wild fauna and flora*. Washington, D.C.

Agreement concerning trade in endangered biota and their products. Appendices list scientific and common names of protected species and the nations that agreed to the Convention.

44. IUCN Conservation Monitoring Centre. 1986. *1986 IUCN red list of threatened animals*. Gland, Switzerland: International Union for Conservation of Nature and Natural Resources.

Lists of species includes 3117 animal taxa: vertebrate and invertebrate.

45. IUCN Conservation Monitoring Centre. 1988. *1988 IUCN Red List of threatened animals*. Cambridge, U.K., and Gland, Switzerland: IUCN.

46. Jackson, P. 1987. Animals and plants on the verge of extinction. *IUCN Bulletin* (Special Report) 18(1-3): 1-16.

    Both author and sponsor are active contributors.

47. Jenkins, R.E., Jr. 1985. The identification, acquisition, and preservation of land as a species conservation strategy. In *Animal extinctions: what everyone should know,* ed., R.J. Hoage, 129-145. Washington, D.C.: Smithsonian Institution Press.

    Article examines the value of state surveys as a means of identifying endangered species and the habitat they need. Cites the example of the Indiana bat and the Texas blind salamander. Volume contains 12 papers by well known authors. While not new material, the text is authoritative.

48. Jones and Stokes Associates, Preprs. 1987. *Sliding toward extinction: the state of California's natural heritage.* San Francisco: Nature Conservancy.

    Not seen.

49. Kallman, H., ed. 1987. *Restoring America's wildlife: the first 50 years of the Federal Aid in Wildlife Restoration (Pittman-Robertson) Act.* Washington, D.C.: Supt. of Documents.

    Authorized by the Fish and Wildlife Service of the U.S. Department of the Interior, this 394-page book tells the history in glowing terms of wildlife management techniques begun in 1937 under the Federal Aid in Wildlife Restoration Act. Well known writers contribute sections on prized game species. State and territory agencies contributed to the publication. See also item 97.

50. Kaufman, L., and K. Mallory. 1986. *The Last Extinction.* Based on a public lecture series, "Extinction: saving the sinking ark" held in Boston at the New England Aquarium during the fall, 1984. Cambridge, Mass.: New England Aquarium and the MIT Press.

    Authors believe a major period of extinction has begun, mostly occurring in the tropics where two-thirds of the extinctions will take place. They note that extinctions are not natural and can be prevented. They describe 25 vertebrate species and sub-species that have become extinct in North America since 1950.

51. Kiew, B.H. 1985. Malaysia's 10 most endangered animals, plants and areas. *Malayan Naturalist* 38(4) May: 2-3.

    A brief paper that lists the author's selection of the most, but not the only, endangered Malaysian animals. Prepared in response to the IUCN list of the 12 most threatened that the author felt was inadequate for Malaysia.

52. Lande, R. 1988. Genetics and demography in biological conservation. *Science* 241(4872)Sept 16: 1455-1460.

While information about two bird species is included, this article is more important as a theoretical examination of the approaches to the determination of viable populations, comparing the importance of population genetics and demography. Demographic factors may be of more immediate importance in planning for species survival of wild populations but author argues "a balance of both are applicable to species in natural environments." Article is a good review of the literature and the accompanying bibliography is essential for this discussion.

53. Leopold, A.S. 1986. *Wild California: vanishing lands, vanishing wildlife.* Contributions by R.F. Dasmann. Photographs by T.A. Blake. Los Angeles: Univ. of California Press.

The author was a respected early researcher and the son of Aldo Leopold, a pioneer in wildlife management.

54. Majumdar, S.K, F.J. Brenner, and A.F. Rhodes, eds. 1986. *Endangered and threatened species programs in Pennsylvania and other states: causes, issues and management.* Harrisburg, Pennsylvania: Pennsylvania Academy of Science.

55. Matthews, J., and D. Lowe, eds. 1989. *Official World Wildlife Fund guide to endangered species.* 2 vols. Washington, D.C.: Beacham Publishing.

Not seen. The publication was anticipated for 1988. *Books in Print* include the title in their 1989 edition. Its sponsorship promises a most complete and authoritative publication.

56. Matthiessen, P. 1987. *Wildlife in America.* Revised and updated edition. New York: Viking Penguin.

An updated edition of the history of American wildlife from the time of the European arrival in the New World to the present that includes every well known case of an animal that is endangered or recently extinct. A sympathetic tale of the animals that vanished, or animals so rare that tardy protection barely saved them from extinction, and of the early efforts and those of the present day to preserve wildlife. A useful bibliographic survey of American natural history is imbedded in the text. Appendices list protected species by geography and common name and a chronology of milestones in legislative conservation activity. A bibliography and an index add to the volume's research usefulness.

57. Mittermeier, R.A. 1981. Brunei protects its wildlife. *Oryx* 16(1)May: 67-70.

Effective laws are preserving the proboscis monkey, clouded leopard, dugong, sea turtles, hornbills, and the earless monitor lizard.

58. Mowat, F. 1984. *Sea of slaughter.* Boston and New York: Atlantic Monthly Press. Repr. 1986, Bantam.

A readable 405-page account of the killing of birds, fish, buffalo, and marine mammals in the North Atlantic Ocean and the inland waterways of eastern Canada and New England. The author is well known as a champion of the environment. Bibliography.

59. Mukherjee, A.K. 1982. *Endangered animals of India.* Calcutta: Zoological Survey of India.

    A detailed description of 4 reptiles, 10 birds, and 29 mammals accompanied by illustrations and maps showing distribution. Conservation policy is also described. A useful table of Indian wild life reserves indicates extent of government policy.

60. Munton, P. 1984. Species in the 1980's. Let's be specific. SSC examines the state of the ark. *IUCN Bulletin* 15(7-9): 83-88.

    Not seen.

61. Murray, M., and S. Oldfield. 1984. Species in the 1980s. Choosing a top ten. *IUCN Bulletin* 15 (7-9): 79-82.

    Not seen but IUCN reports are usually valuable.

62. National Wildlife Federation. 1988. Twentieth Environmental Quality Index. *National Wildlife* 26(2) Feb-Mar: 38-46.

    Published annually in *National Wildlife*, the Federation compiles a checklist of environmental criteria and evaluates the year's events and conditions as an "indicator" of the status of the environment. This issue highlights the California condor, the dusky seaside sparrow, the bald eagle and the American alligator.

63. Nilsson, G. With contributions by M. Gean, J. Buckley, and B. Greveman. 1983. *The endangered species handbook.* Washington, D.C.: Animal Welfare Institute.

    A very comprehensive overview of the status of the movement to save endangered species: organizations, publications, programs, law and legislation. Appendix includes tables showing mammals, birds, reptiles, and amphibia that have become extinct since 1600. Another table lists endangered and threatened species according the many organizations that create such lists: IUCN, CITES, U.S. Fish and Wildlife Service, and the states that issue their own lists.

64. Norden, A.W., D.C. Forester, and G.H. Fenwick. 1984. *Threatened and endangered plants and animals of Maryland.* Special Publication/ Maryland Natural Heritage Program: 84-I. Annapolis, Md.: Department of Natural Resources.

    Proceedings of a symposium held September 3-4, 1981 at Towson State University, Towson, Md.

65. Norris, R. 1988. Data for diversity. *Nature Conservancy Magazine* 38(1)Jan/Feb: 4-10.

Describes criteria selected by the Nature Conservancy to monitor biological status of Latin America and the Caribbean.

66. Norton, B.G., ed. 1986, 1988 (pap). *The Preservation of species. The value of biological diversity.* Written under the auspices of the Center for Philosophy and Public Policy, University of Maryland. Princeton: Princeton Univ. Press.

Few individual species are mentioned. Primarily a theoretical work addressing basic definitions of the value of species, criteria for preservation priorities, philosophical and religious attitudes, and management concerns.

67. Nye, P.E. 1984. Managing the bottom 12 percent: New York's endangered species program. *Conservationist* (New York) 39(3)Nov-Dec: 26-35.

The role of New York's Department of Conservation in protection through a program of listing species in several categories of endangerment and promoting programs for restoration. Although written as a public relations document, useful information is provided, in particular, a table of endangered, threatened, and animal species of special concern.

68. Olney, P.J.S., ed. 1986. *International zoo yearbook 24/25*. London: Zoological Society of London.

Twenty-eight papers on the special subject of breeding endangered species in captivity presented at a conference in 1984.

69. Peters, R.L., and J.D.S. Darling. 1985. The greenhouse effect and nature reserves; global warming would diminish biological diversity by causing extinctions among reserve species. *Bioscience* 35(11)Dec: 707-718.

Predicts that surviving species in the next century will be isolated in small patches of their original habitat. Paper details effects of climatic changes on biological communities and reserves.

70. Peterson, M.A., and D. Adler. 1982. *The effects of air pollution and acid rain on fish, wildlife, and their habitats: critical habitats of threatened and endangered species.* U.S. Fish and Wildlife Service, Office of Biological Service, Report No. 11. Washington, D.C.

Not seen.

71. Rana, B.D., and R. Advani. 1983. Endangered wildlife in Rajasthan desert. *Tigerpaper* 10(1): 5-7.

The journal is one of the best sources for news of Asian conditions but is only available at some research libraries and the Smithsonian.

72. Raven, P.H. 1987. We're killing our world: preservation of biological diversity. *Vital Speeches the Day* 53(May 15): 472-478.

Delivered as Keynote Address to the AAAS annual meeting, February 1987. Describes the birds declining on Barro Colorado Island and in Hawaii where only 28 species of land birds (out of 88 known) are left. Prophesizes regional climatic problems and loss of biological diversity if deforestation continues. It is the tropical areas where the greatest concentration of biological diversity exists. Impassioned description of dangers to the planet from disregarding tropical forests where the third world lives under great population pressures.

73. Reed, N.P., and D. Drabelle. 1984. *The United States Fish and Wildlife Service*. Boulder, CO.: Westview Press.

    A history of the F&WS, the federal agency that manages the national wildlife refuge system, protects endangered species, and conducts fish and wildlife research. Useful also for the discussion of major theoretical conflicts during the agency's existence.

74. Rolston, H., III. 1985. Duties to endangered species. *Bioscience* 35(11)Dec: 718-726.

    Urges an ethic to preserve species that balances biology and human needs. Explores controversies involved in extensiveness of preservation activity, e.g., preserve vermin?

75. Roots, C. 1987. *Endangered species: Canada's disappearing wildlife*. Illus. C. Godkin. Markham, Ont.: Fitzhenry & Whiteside.

    Primarily an identifying list of 101 species considered lost or those that may soon disappear; their description and causes of decline.

76. Saharia, V.B. 1982. *Wildlife in India*. Dehra Dun, India: Natraj Publishers.

    Prepared on the occasion of a 1981 CITES meeting in New Delhi. A chapter entitled "Special Projects for Endangered Species" is a useful discussion of management and natural history of six species of special concern. Each receives a full case history. As one animal, the tiger, is a threat to villagers' lives, the statement of government objectives is pertinent.

77. Sarkar, S.U. 1985. Endangered wildlife of the coastal zone of Bangladesh. *Coastal Zone* 1:1110-1120.

    Not seen. Author is prominent in this field.

78. Sarker, S.U., and N.J. Sarker. 1983. Endangered wildlife of Bangladesh. *Tigerpaper* 10(2): 26-28.

79. Schaefer, J. 1985. The extinct: losses and lessons. *Kansas Wildlife* 42(2): 26-30.

80. Shaffer, M.L. 1981. Minimum population sizes for species conservation. *Bioscience* 31(2)Feb: 131-134.

Development pressures on land require that nature reserves be used as effectively as possible to preserve species' populations. Author suggests definition of 'minimum viable population' and the factors that determine population size. An important article that sets out theoretical approaches for consideration in determining the minimum size population necessary for continuation of a species.

81. Shaffer, M.L., and F.B.Samson. 1985. Population size and extinction: a note on determining critical population sizes. *American Naturalist* 125(1)Jan: 144-151.

Scholarly critique of theories and models intended to assess size of a stable population and land area. Authors analyze several models designed to predict stochastic extinction, comparing them to Shaffer's computer simulated model for determining size of population and extinction in the case of the grizzlies of Yellowstone.

82. Skoog, P.J. 1982. *Highways and endangered wildlife in Florida: impacts and recommendations*. Final Report. Tallahassee: Federal Highway Administration, Floridal Game and Freshwater Fish Commission. FHWA/FL/BMR-84/272.

Designed to advise the Florida DOT on techniques to avoid or lessen impacts on endangered species, the text lists those that are endangered and threatened. Data on life histories, habitat needs, and vulnerability to disturbance assembled for recommendations. Species' descriptions are brief. Manual is primarily concerned with legal status of species and environmental dangers.

83. Soule, M.E. 1985. What is conservation biology?*Bioscience* 35(11)Dec: 727-734.

An overview of a new discipline that addresses the issues of disturbed species and ecosystems.

84a. ______. ed. 1986. *Conservation biology: the science of scarcity and diversity*. Sunderland, MA: Sinauer Assoc.

The title represents a new academic discipline that studies the abundance and decline of species. The book is a collection of theoretical and research papers that includes many case histories of endangered species and their habitats.

84b. ______. 1987. *Viable populations for conservation*. Cambridge, (U.K.) New York, Sydney: Cambridge Univ. Press (U.K.)

Nine contributed chapters concerned with a current basic question in conservation biology, i.e., "what size population is needed for survival of a species in a state that maintains its vigor and its potential for evolutionary adaptation."

85. Steadman, D.W., and S. Zousmer. 1988. *Galapagos: discovery on Darwin's*

islands. Washington, D.C.: Smithsonian Institution Press.

Against the background of the history and importance of Darwin's observations of these islands, the current residents are described accompanied by Lee Steadman's drawings.

86. Stermer, D. 1981. *Vanishing creatures, a series of portraits*. Introduction by Robert Redford. Berkeley, Calif.: Lancaster-Miller Publishers.

Twenty-six artistic illustrations accompanied by a very brief analysis of causes of their decline. Appendices are particularly useful: lists of endangered species as determined by the IUCN, U.S. Government classification, and private sources.

87. Stone, C.P., and D.B. Stone, eds. 1988. *Conservation biology in Hawaii*. University of Hawaii National Park Resources Studies Unit. Honolulu: Univ. of Hawaii Press.

Collection of papers that stress the diversity of Hawaiian biota, review historical conditions leading to the current status, and the need for immediate conservation efforts.

88. Stuart, S.N., ed. 1986. *Conservation of Cameroon montane forests*. Report of the ICBP Cameroon Montane Forest Survey, Nov. 1983-April 1984. Cambridge, U.K.: International Council for Bird Preservation.

Montane forests have great significance for eight threatened species of birds in Western Cameroon. Each forest bird is described and non-forest birds are listed. Useful information about reptiles, amphibians, and mammals is included.

89. Tikader, B.K. 1983. *Threatened animals of India*. Calcutta: Zoological Survey of India.

Presents full page discussions of 18 mammals, 47 Birds, 15 reptiles and amphibians selected from the Indian Wild Life Act. Information on distribution, habits and habitat, and status is accompanied by a color illustration.

90. Tohme, G., and H. Tohme. 1981. Extinct and disappearing animals in Lebanon. *Biology International* 3:10-14.

Not seen. The publication is a newsletter and difficult to find but the topic is a unique one.

91. Torres, J., S. Bissell, G. Craig, W. Graul, and D. Langlois. 1986. *Wildlife in danger: the status of Colorado's threatened or endangered birds, fish and mammals*. Colorado Division of Wildlife: Denver.

A 31-page booklet that may be difficult to get in other than Colorado libraries or from the state's Division of Wildlife. For information on Colorado mammals see **Mammals** Chapter, Murray, J.A.

92. Total Environment Centre, Endangered Species Committee (Australia).

1983. *Our wildlife in peril.* French Forest, N.S.W.(Australia): Reed.

Describes Australia's endangered animals.

93. Tubbs, C. 1985. The decline and present status of the English lowland heaths and their vertebrates. *Focus on Nature Conservation* 11:1-20.

94. U.S. Congress. House. Committee on Merchant Marine and Fisheries. Subcommittee on Fisheries and Wildlife Conservation and Environment. 1987. *Endangered Species Act reauthorization: hearings. on H.R. 1467.* 100th Cong., 1st sess. March 17,1987

A 725-page volume that includes statements by legislators, lists of and statements by interested individuals and organizations, and exhibits presented at the hearings. They may consist of useful articles and reports. Marine turtles were the most frequently mentioned.

95. U.S. Fish and Wildlife Service. 1984, 1986. *Endangered and threatened wildlife and plants.* 50 CFR 17.11 and 17.12; Title 50-Part 17, Subpart B-Lists. Washington, D.C.: U.S. Government Printing Office.

Official list names all species of wildlife that have been determined by the Service to be endangered or threatened. Look for updates.

96. U.S. Fish and Wildlife Service. 1981. *Endangered means there's still time.* Washington, D.C.: Fish and Wildlife Service.

A 33-page booklet that is an excellent summary to introduce newcomers to the role of the United States government in protecting wildlife. Very brief descriptions of selected endangered animals are included.

97. U.S. Fish and Wildlife Service. 1987. *Restoring America's Wildlife.* Ed., H. Kallman. Washington, D.C.: Supt. of Documents.

See also item 49.

98. U.S. Fish and Wildlife Service, Office of Endangered Species. 1983. *Liaison conservation directory for endangered and threatened species.* 5th Ed. Washington, D.C.: U.S. Fish and Wildlife Service.

A directory of every government agency that has anything to do with the conservation of endangered species. This is the latest edition but be aware of out-of-date facts.

99. Vandenbeld, J. 1988. *Nature of Australia: a portrait of the island continent.* New York: Facts on file. Published in association with the Australian Broadcasting Association.

Attractive, comprehensive view of Australian nature extending to Australasian islands. No indication of endangerment or rarity in index, but such status is mentioned in species' descriptions as in the section about the dibbler. Reader needs to identify endangered species elsewhere before turning to this book for information about the animal.

100. Veziroglu, T.N., and B.N. Kumar, eds. 1984. *Endangered species in Guyana and the Caribbean: The biosphere: problems and solutions*. Proceedings of the Miami International Symposium on the Biosphere 23-24 April 1984, Miami Beach, Florida. New York: Elsevier.

Not seen.

101. Wallace, D.R. 1987. *Life in the balance*. New York: Harcourt Brace Jovanovich.

Based on the highly acclaimed TV series produced by the National Audubon Society describing the state of our environment, the species that are in danger because of man, and man's efforts to recover them. Explores Earth's ecosystems – skies, grasslands, forests, deserts, etc.—and offers dramatic portraits of some of the world's rarest creatures along the way.

102. Warhol, A. (Color illustrator), and K. Benirschke.(Text). 1986. *Vanishing animals*. New York:Springer-Verlag.

The zoology of 15 endangered species accompanied by a full-page color illustration of each prepared by Warhol as a silkscreen over collage. Interesting for its art. Text consists of brief summaries.

103. Warner, R.E., and K.M. Hendrix, eds. 1984. *California riparian systems: ecology, conservation and productive management*. Berkeley, Los Angeles, and London: Univ. of California Press.

Collected papers that are the result of studies requested by the State for management of areas that are under population pressures. Several specifically relevant papers are indexed throughout the present volume.

104. Westing, A.H., and C.E. Westing. 1981. Endangered species and habitats of Viet Nam. *Environmental Conservation* 8(1): 59-62.

During a 1980 visit to Viet Nam the authors explored general habitats and current status of a number of endangered species.

105. White, C.P. 1982. *Endangered and threatened wildlife of the Chesapeake Bay Region: Delaware, Maryland, and Virginia*. Centerville, Md.: Tidewater Publishers.

106. Williams, G.R., and D.R. Given, eds. 1981. *The Red Data Book of New Zealand. Rare and endangered species of endemic terrestrial vertebrates and vascular plants*. Wellington: Nature Conservation Council.

Not seen.

107. Wilson, E.O., and F.M. Peter, eds. 1988. *Biodiversity*. Washington, D.C.: National Academy Press.

A collection of papers addressing all major conservation issues. Volume is the result of a large conference held in 1986 and sponsored by the National Academy of Science and the Smithsonian Institution. Scientists see biodiversity as the fabric of nature, interconnecting webs of animal and plant dependencies that are the basis of human life. The disappearance of species is the concern that prompted the conference. All the papers are gripping.

108. Wilson, J.H., comp. and ed. 1984. *Rare and endangered species of Missouri*. Jefferson City, MO: Missouri Department of Conservation.

Defines "endangered, rare, status undetermined, and possibly extirpated." Lists 22 mammals, 24 birds, 14 amphibians and reptiles, and 36 fish. Distribution shown on maps. Nesting sites and status are described.

109. Wolf, E. 1987. On the brink of extinction: conserving the diversity of life. *World watch Paper* 78:1-54.

110. Wood, D.A. 1981. *Endangered species: concepts, principles, and programs: a bibliography*. Tallahassee: Florida Game and Fresh Water Fish Commission.

The literature up to 1980.

111. Wood, D.A., Comp. 1984. *Official lists of endangered and potentially endangered fauna and flora in Florida*. Tallahassee: Florida Game and Fresh Water Fish Commission.

A small (16 pages) but useful list. Florida is the only sub-tropical state in the United States mainland and its fauna are of particular interest.

112. Woolfenden, G.E. 1983. *Rare, threatened, and endangered vertebrates of southwest Florida and potential OCS (Outer Continental Shelf) activity impacts*. FWS/OBS 82/03. Distributed by NTIS.PB83-250811. Washington, D.C.: U.S. Fish and Wildlife Service, Office of Biological Services, and Bureau of Land Management, Branch of Offshore Studies.

Compilation of all species so designated or considered for listing by various federal, state and private institutions and organizations. Identifies vertebrates of southwest Florida that could be affected by OCS development. Each vertebrate is listed in the habitat it occupies. Information about reproduction, feeding, and population estimates is given. Useful tables with common and scientific names are included.

113. Yaffee, S.L. 1982. *Prohibitive policy*. MIT Studies in American politics and public policy, No. 9. Cambridge, MA and London, England: MIT Press.

Primarily concerned with government regulation programs affecting 1973 Endangered Species Act. A brief synopsis of several case studies is included.

114. Yom-Tov, Y., and E. Tchernov, eds. 1988. *The zoogeography of Israel: the distribution and abundance at a zoogeographical crossroad*. Monographiae Biologicae, vol. 62. Norwell, MA: Kluwer.

Not seen.

MAMMALS

* Aiken, S.R., and C.H. Leigh. 1985. On the declining fauna of peninsular Malaysia in the postcolonial period. *Ambio* 14(1): 15-23.

Elephants, tigers, and the Sumatran rhinoceros are among the mammals facing extinction. See also **General** Chapter, item # 1.

115. Ali, R. 1985. An overview of the status and distribution of the lion-tailed macaque. *Monographs in Primatology* 7:13-25.

* Anadu, P.A. 1987. Progress in the conservation of Nigeria's wildlife. *Biological Conservation* 41(4): 237-251.

Described in **General** Chapter, item # 3.

116. Aveling, R., and A. Mitchell. 1982. Is rehabilitating Orang Utans [sp] worth while? *Oryx* 16(3)Feb: 263-271.

Preservation is related to habitat and management.

117. Ayres, J.M., R.A. Mittermeier, and I.D. Constable. 1982. Brazilian tamarins on the way to extinction? *Oryx* 16(4)June: 329-333.

Wide ranging animals have better chance for survival than those with restricted ranges whose extinction is threatened when their small habitat is disturbed by development.

118. Baker, M.L. 1987. *Whales, dolphins, and porpoises of the world.* Garden City, NY: Doubleday.

Good source of information about all types of whales with descriptions and names of each species of cetacean order.

119. Balakrishnan, M. 1984. The larger mammals and their endangered habitats in the Silent Valley Forests of south India. *Biological Conservation* 29(3): 277-286.

During a study of the habitat and population of large mammals to assess the likely impact on fauna by a proposed hydroelectric dam, useful information about the animals was gathered.

120. Balcomb, K.C., III. 1982. *Marine birds and mammals of Puget Sound.* Washington Sea Grant Program. Seattle: Univ. of Washington Press.

The blue and humpback whales of this area are described. Also cited in **Birds** chapter.

121. Banks, V. 1988. The red wolf gets a second chance to live. *Smithsonian* 18(12)March: 101-107.

One of the most beleaguered mammals in the U.S. Seven red wolves raised in captivity were released in a North Carolina refuge in a historic attempt to restore a self-sustaining population.

122. Barnes, V.G., Jr., R.M. Anthony, K.A. Fagerstone, and J. Evans. 1985. Hazards to grizzly bears *Ursus arctos* of strychnine baiting for pocket gopher (Thomomys talpoides) control. *Wildlife Society Bulletin* 13(4)Winter: 552-558.

Forests have planned reforestation units that are threatened by pocket gophers who eat conifer seedlings. FWS questioned threat to grizzlies of strychnine treated grain set out for pocket gophers.

123. Bauer, E.A. 1988. *Erwin Bauer's predators of North America. [Predators of North America]*. Latham, New York: Grolier. [Distributed by Stackpole Books, Harrisburg, PA]

A very readable and anecdotal account of endangered, rare, threatened, and protected mammals. Written in an engaging style ringing with the vividness of observing animals in the wild and accompanied by natural setting photography.

124. Beck, B.B., and C. Wemmer, eds. 1983. *Biology and management of an extinct species—Pere David's deer*. Park Ridge, NJ: Noyes Publications

A series of papers about the captive breeding programs for the Pere David's deer.

125. Belden, R.D., W.B. Frankenberger, R.T. McBride, and S.T. Schwikert. 1988. Panther habitat use in southern Florida. *Journal of Wildlife Management* 52(4)Oct: 660-663.

Authors studied captured panthers in both swamp and hammock forests. Concluded that south Florida can support 30-40 panthers.

126. Bell, D.J., J. Hoth, A. Velazquez, F.J. Romero, L. Leon and M. Aranda. 1985. A survey of the distribution of the volcano rabbit, *Romerolagus diazi*: an endangered Mexican endemic. *Dodo* No. 22:42-48.

* Bendiner, R. 1981. *The fall of the wild, the rise of the zoo*. New York: E.P. Dutton.

Breeding endangered animals in zoos. Appendix is a table of endangered species including mammals. See also **General** Chapter, item 10.

127. Benirschke, K. ed. 1986. *Primates: the road to self-sustaining populations*. New York: Springer Verlag.

Compendium of 74 papers on primates that were presented at a 1985 conference. Combines information on abundance and status of primate species with conservation strategy. Scholarly.

* Berger, T.J., A.M. Neuner, and S.R. Edwards. 1981. *Directory of Federally Controlled Species*. Lawrence, Kansas: Association of Systematics Collections.

Cited in **General** Chapter as item 11.

128. Bergman, C.A. 1982. Endangered species: yesterday's caribou hang on in northern Idaho. *Audubon* 84(5)Sept: 110-114.

129. Block, R. 1981. Michigan's endangered species program. *Endangered Species Technical Bulletin* 6(6): 4-6.

News of this state's program for Indiana bats, pine marten, and the timber wolf.

130. Bohlen, J.T. 1988. Demystifying the right whale. *Defenders* 63(5)Sept-Oct: 10-19.

Report of a study of migration patterns in order to identify the right whale's habitat. Following that, protection efforts for the habitat can begin.

131. Bonner, W.N. 1982. *Seals and man: a study of interactions*. A Washington Sea Grant publication. Seattle: Univ. of Washington Press.

A study of many seals, some endangered and some not. Shows distribution, breeding patterns, harvesting history by man, nutrition needs, migratory patterns and history of hunting.

132. Booth, W. 1988. Reintroducing a political animal. *Science* 241(4862)Jul 8: 156-158.

Government biologists attempting to move the threatened sea otters to a more protected area are thwarted by otters returning to familiar territory.

133. _______. 1988. Chimps endangered, research reprieved. *Science* 242(4883) Dec 2: 1240.

U.S. Fish and Wildlife Service reclassified wild chimps as "endangered" but those living in research facilities remain "threatened," a policy favored by Jane Goodall. A single column news item describing a current controversy.

134. Borner, M. 1981. Black rhino disaster in Tanzania. *Oryx* 16(1)May: 59-66.

Poaching is about to exterminate the black rhino in northern Tanzania. Describes efforts to halt poaching.

135. Borner, M., C.D. FitzGibbon, MO. Borner, T.M. Caro, W. K. Lindsay, D.A. Collins, and M.E. Holt. 1987. The decline of the Serengeti Thomson's gazelle. *Oecologia* (Berlin) 73(1)Aug: 32-40.

Censuses of Thomson's gazelle showing 2/3ds population decline over the past 13 years. Exploration of causes point to predators and disease.

* Boyd, J.M. 1984. Nature conservation. *Proceedings of the Royal Society of Edinburgh,* Section B-Biological Sciences. 84: 295-335.

The shearwater manx, St. Kilda mouse, and the grey seal are among the mammals described. Cited in **General** Chapter as item 12.

136. Brack, V., Jr., A.M. Wilkinson, and R.E. Mumford. 1984. Hibernacula of the endangered Indiana bat in Indiana. *Proceedings of the Indiana Academy of Science* 92:463-468.

137. Brady, J.R., and D.S. Maehr. 1985. Distribution of black bears in Florida. *Florida Field Naturalist* 13(1): 1-7.

138. Braham, H.W. 1984. The bowhead whale, *Balaena mysticetus. Marine Fisheries Review* 46(4): 45-53.

Entire issue devoted to endangered whales.

139. ______, and D.W. Rice. 1984. The right whale, *Balaena glacialis. Marine Fisheries Review* 46(4): 38-44.

140. Brannon, R.D., R.D. Mace, and A.R. Drood. 1988. Grizzly bear mortality in the northern continental divide ecosystem, Montana. *Wildlife Society Bulletin* 16(3): 262-269.

Authors studied causes of deaths to discover mortality patterns that would aid management programs. Argues for a quota system to regulate hunter harvest that would arrest increases in deaths.

141. Brown, D.E., and D.M. Gish, eds. 1983. *The wolf in the southwest: the making of an endangered species.* Tucson: Univ. of Arizona Press.

Contributing authors are R.T. McBride, G.L. Nunley, and J.F. Scudday.

142. Brown, D.E. 1985. *The grizzly in the southwest: documentary of an extinction.* Norman: Univ. of Oklahoma Press.

Decline of the grizzly came about because ranchers and others considered the bear a dangerous predator. Author argues that the grizzly could have been saved with proper management but the will to do so has came too late.

143. Brueggeman, J.J., and G.A. Green. 1986. Endangered whale surveys of the northwestern Gulf of Alaska and southeastern Bering Sea. *Cetus* 6(2): 20-21.

144. Burke, R.L., and S.R. Humphrey. 1987. Rarity as a criterion for endangerment in Florida's fauna. *Oryx* 6(2): 97-102.

145. Burton, J.A., assisted by V.G. Burton. 1988. *Collins guide to the rare mammals of the world.* Lexington, Mass: S. Greene Press; New York: Distributed by Viking Penguin.

Not seen. Includes a foreword by Peter Scott, an active conservationist. Illustrated by B. Pearson.

* Byrne, G. 1988. Strengthened endangered species act passes. *Science* 242(4876)Oct 14: 190.

Cited as item 14 in **General.**

146. ______. 1988. California, here they go. *Science* 242(4885)Dec 23: 1639.

A brief update concerning the number of vertebrate classes in California and the awareness of their decline. Describes the California Extinction Index developed by the California Nature Conservancy.

* Cadieux, C. 1981. *These are the endangered.* Drawings by Bob Hines. Washington, D.C.: Stone Wall Press; Brattleboro, Vt.: Distributed by Stephen Green Press.

Well known endangered mammal species are noted. See item 15 in **General** Chapter.

147. Cahn, R., and P. Cahn. 1985. Saved but threatened. *Audubon* 87(3)May: 48-51.

Lists 12 most threatened animals as identified by the IUCN.

148. Campbell, R.R. 1984. Rare and endangered fishes of Canada: The Committee on the Status of Endangered Wildlife in Canada (COSEWIC) Fish and Marine Mammals Subcommittee. *Canadian Field-Naturalist* 98(1)Jan-Mar: 71-74.

Describes the function of COSEWIC (the Canadian "endangered species list") and provides a list of fishes and marine mammals that are on the list as well as a table showing those species being considered. A useful starting point for identifying endangered fish and marine species. Cited also in **Fishes** chapter.

149. ______. 1985. Rare and endangered fishes and marine mammals of Canada: COSEWIC Fish and Marine Mammals Subcommittee status report: 2. *Canadian Field-Naturalist* 99(3): 404-408.

Follow-up to a series of status reports published as articles in the *Canadian Field-Naturalist* (1984,98(1): 63-133) that analyzed the population of candidates for listing on COSEWIC, the Canadian endangered species list. Author introduces seven status reports of individual species that follow. Article includes tables of fish and marine mammal species that have been assigned COSEWIC status and those that are of interest to COSEWIC.

150. _______. 1987. Rare and endangered fishes and marine mammals of Canada: COSEWIC Fish and Marine Mammal Subcommittee status reports, 3. *Canadian Field-Naturalist* 101(2): 165-170.

Follow-up to item 149.

151. Carey, J. 1987. Who's afraid of the big bad wolf? *National Wildlife* 25(Aug-Sept): 4-12.

Defends the value of reintroducing the wolf to Yellowstone.

152. Caro, T.M., and D.A. Collins. 1987. Ecological characteristics of territories of male cheetahs. *Journal of Zoology* 211:89-105.

Investigations of ecological determinants causing male cheetahs to establish territories in the Serengeti: rainfall, vegetation, prey, etc.

153. Carroll, J.B. 1983-84. The conservation and wild status of the Rodriques fruit bat, *Pteropus rodricensis*. *Myotis* 21-22:148-154.

154. Cauble, C. 1986. Return of the native. The Park Service calls for reintroduction of Yellowstone's missing predator... the wolf. *National Parks* 60(7-8): 24-29.

The "official" view.

155. Ceballos-G, G. 1985. The importance of riparian habitats for the conservation of endangered mammals in Mexico. *Rocky Mountain Forest and Range Experiment Station, General Technical Report,* U.S. Forest Service, Fort Collins, CO. pp. 96-100.

156. Chaplin, G. 1985. El tigre hombre. (Jaguars in Belize.) *Audubon* 87(4)July: 40-44.

157. Chase, A. 1983. The last bears of Yellowstone. *Atlantic Monthly* 251(2) Feb: 63-67,70-73.

Disputes the Park Service's population estimates for black bears and for grizzlies, and critical of the information distributed by the Service. Provides an overview of park policies and controversies concerning bear management since 1967.

158. Chehebar, C.E. 1985. A survey of the southern river otter *Lutra provocax* Thomas in Nahuel Huapi National Park, Argentina. *Biological Conservation* 32(4): 299-307.

This amphibious mustelid is listed under "indeterminate status" in the IUCN Mammal Red Data Book (1982), as "endangered" in the Argentine National Wildlife List and the Patagonian Vertebrate Inventory. Argues the importance of the park in the conservation of this species.

159. Clark, B. 1988. On the trail of Israel's lost desert cat. *International Wildlife* 18(Jan-Feb): 12-15.

Thought to have become extinct, several have been sighted and are being studied.

160. Clark, D.R., F.M. Bagley, and W. Johnson. 1988. Northern Alabama colonies of the endangered grey bat, *Myotis grisescens*: organochlorine contamination and mortality. *Biological Conservation* 43(3): 213-227.

161. Clark, M.K., ed. 1987. *Endangered, threatened and rare fauna of North Carolina*. Part 1, *A reevaluation of the mammals*. Occasional papers of the

North Carolina Biological Survey 1987. Raleigh: North Carolina State Museum of National Sciences.

162. Clark, T.W. 1986. Black-footed ferret: ghost of the prairie. *Animal Kingdom* 89(3)May-June: 38-48.

Black-footed ferret thought to have become extinct after 1974 but discovery of a wild colony in Wyoming brought opportunity for research. Article describes animal's behavior.

163. Clark, T.W., S.C. Forrest, L. Richardson, D.E.Casey and T.M. Campbell III. 1986. Description and history of the Meeteetse black-footed ferret environment. *Great Basin Naturalist Memoir* 8: 72-84.

More scholarly account than previous reference. Entire issue devoted to black-footed ferret.

164. Clawson, R.L. 1987. Indiana bats: down for the count. *Endangered Species Technical Bulletin* 12(9): 9-11.

165. Coe, M. 1982. The bigger they are... *Oryx* 16(3) Feb: 225-228.

Large body size may enable elephant avoid predators and to cover large areas in search of food. Large body size also requires large amounts of food meaning continuous feeding and large land areas to sustain adequate food needs. Long gestation periods and dependency of young contribute to vulnerability. Only a small harvest should be taken but poachers with automatic weapons may cause extinction. Author explores periods of extinction and relationship to hunting.

* Coffin, B., and L. Pfanmuller, eds. 1988. *Minnesota's endangered flora and fauna*. Minn.: Univ. of Minnesota Press.

Elk, wolves, wolverines, caribou, and other mammals are among the fauna described. See item 17 in the **General** Chapter.

166. Cohn, J.P. 1986. Surprising cheetah genetics. *Bioscience* 36(6)June: 358-362.

Report of an in-depth study revealing the lack of genetic diversity in cheetahs regardless of their geographic origin.

167. Cohn, J.P. 1988. Captive breeding for conservation. *Bioscience* 38(5)May: 312-316.

The author states that "despite success stories, this strategy remains controversial within the conservation community." Reviews the major arguments pro and con captive breeding. The Arabian oryx, Przewalski's horse, and Pere David's deer have been successfully bred in zoos and returned to nature. Those opposed to captive breeding argue that money would be better spent on preserving habitats rather than a few individuals. Describes ISIS, the International Species Inventory System.

168. Conner, D.K., and R. O'Dell. 1988. The tightening net of marine plastics pollution: Strategies for intervention. *Environment* (Washington, D.C.) 30(1)Jan-Feb: 16-20, 33-36.

Describes role of plastic refuse in killing marine life, e.g. green turtles and manatees. Also cited in **Amphibians and Reptiles** chapter.

169. Conniff, R. 1987. When the music in our parlors brought death to darkest Africa. *Audubon* 89(7) Jul: 76-93.

Elephants and ivory for piano keys.

170. Contreras, G.P., and K.E. Evans, eds. 1986. *Proceedings of the grizzly bear habitat symposium, April 30-May 2, 1985*. Technical Report, International Research Station No. 207. Missoula, Montana and Ogden, Utah: U.S. Department of Agriculture, Forest Service.

Thirty-three papers and 3 abstracts provide current information on grizzly bear habitat delineation and management.

* Convention on International Trade in Endangered Species (CITES). 1981. *Identification manual*. Vol. 1, *Mammalia* 5 vols. Gland, Switzerland: Convention on International Trade in Endangered Species of Wild Fauna and Flora.

Manual excellent for identifying species of concern and a descriptive profile of each is included. Common and scientific name, characteristics of appearance and geographic range are noted for each. Also cited in **General** Chapter as item 19.

171. Corbet, B.B., and J.D. Hill. 1986. *A world list of mammalian species*. 2nd ed. New York: Facts on File.

Entries of endangered mammals are highlighted.

* Cortese, T.J., and B.A. Grosnek, comps. 1987. *Selected publication series of the U.S. Fish and Wildlife Service, 1889-1985*. Washington, D.C.: Fish and Wildlife Service.

Cited in **General** Chapter as item 20.

172. Cox, R. 1987. The Philippine spotted deer and the Visayan warty pig. *Oryx* 6(1): 37-42.

173. Crouse, D. 1988. Up close with gorillas. *National Wildlife* 26(6) Nov-Dec: 5-11.

African tours to see gorillas have become popular and effective as fund-raisers, preventing poaching, and discouraging habitat loss.

174. Day, D. 1987. *The whale war*. San Francisco: Sierra Club.

The effort to save the whales has been written about widely but this organized chronicle of the events documents what has been accomplished in ending whale hunting. The loopholes that permit abuse to continue are noted.

175. De Silva, P.H.D.H. 1987. Cetaceans recorded off Sri Lanka, India, from the Arabian Sea and Gulf, Gulf of Aden, and from the Red Sea. *Journal of the Bombay Natural History Society* 84(3)Dec: 505-525.

Paper explores strandings of 30 whale species during their long migrations between cold seas and warm waters.

176. De-Hua, Y. 1988. Black snub-nosed monkeys in China. *Oryx* 22(1)Jan: 41-43.

Survey of current status.

177. Dearden, P., and C. Hall. 1983. Non-consumptive recreation pressures and the case of the Vancouver Island marmot. *Environmental Conservation* 10(1): 63-66.

Examines reasons for decline in the population of the Vancouver Island marmot, Canada's only land-based endangered endemic species of mammals.

* Diamond, J.M. 1984. "Normal" extinctions of isolated populations. In *Extinctions*, ed., M.N. Nitecki, 193-246. Chicago: Univ. of Chicago Press.

Case studies of mammals and birds on islands, land bridges and archipelagos. See also item 25 in **General** Chapter.

178. Dietz, M.J. 1981. Ecology and social organization of the maned wolf *Chrysocyon brachyurus. Dissertation Abstracts International* 42(12-B): 4706. Ph.D. 1981, Michigan State University.

Nine wolves were studied in Brazil to determine their use of habitat, group behavior, and reproductive patterns. Diet in relation to nature of food and time and season of eating were examined because of ranchers' concerns.

* DiSilvestro, R.L., and A.S. Eno, eds. 1986. *Audubon wildlife report 1986.* New York: National Audubon Society. (Dist. by Academic Press.)

Cited in **General** Chapter as item 26.

* DiSilvestro, R.L., ed. 1987. *Audubon wildlife report 1987*. Orlando: National Audubon Society. (Dist. by Academic Press.)

Cited in **General** Chapter as item 27.

179. Dobson, A.P. 1988. Restoring island ecosystems: the potential of parasites to control introduced mammals. *Conservation Biology* 32(1)Mar: 31-39.

The introduction of small mammals, e.g., rats, cats and mongoose, has decimated many an endemic island population. Study proposes introducing certain parasites to reduce the density of foreign species. Excellent bibliography. See item 488.

180. Doughty, R.W. 1983. *Wildlife and man in Texas*. College Station: Texas A & M University Press.

Argues that decline of fish and game during 19th century was due to overhunting. Conservation measures began in 20th century. Tables of population data of protected species. See item 489.

181. Durrell, G.M. 1983. *Ark on the move*. New York: Coward-McCann.

Bats and lemurs are among the mammals described on Mauritius and Madagascar. Unique preserve on the Isle of Jersey for endangered species. See item 492.

182. Durrell, G.M. 1984, 1977. *Golden bats and pink pigeons*. 2nd ed. New York: Simon and Schuster.

A unique program of conservation. See item 493.

183. Edroma, E.L. 1982. White rhino extinct in Uganda. *Oryx* 16(4)June: 352-355.

Author describes surveys taken since 1925 to prove his point.

* Ehrlich, P.R., and A. Ehrlich. 1981. *Extinction: the causes and consequences of the disappearance of species*. New York: Random House, NY: Ballantine, 1983, c1981. pap.

Cited in **General** Chapter as item 29.

184. Erickson, D.W., and D.A. Hamilton. 1988. Approaches to River Otter restoration in Missouri. In *Transactions of the 53rd North American Wildlife and Natural Resources Conference*. Ed. R.E. McCabe, 404-413. Washington, D.C.: Wildlife Management Institute.

Unregulated otter harvest had all but ended this animal's existence in the range it enjoyed before settlement by Europeans. Description of various strategies to achieve restoration demonstrates the complexity of a restoration project.

185. Evans, P.G.H. 1987. *The natural history of whales and dolphins*. New York: Facts on File.

186. Fa, J.E. 1983. The Barbary macaque – the future. *Oryx* 17(2): 62-67.

Listed in Red Data Book since 1975, the only African non-human primate found north of Sahara and only member of this genus found outside Asia.

187. Fagerstone, K.A., and D.E. Biggins. 1986. Comparison of capture and recapture and visual count indices of prairie dog densities in black-footed ferret habitat. *Great Basin Naturalist Memoir* 8:94-98.

Prairie dog is primary diet for ferret.

188. Fitter, R. 1982. Arabian oryx returns to wild. *Oryx* 16(5)Oct: 406-410.

Dramatic story of the introduction of 10 Arabian oryx born in Phoenix, Arizona into Oman.

189. _______. 1983-84. Rx for an endangered species: Operation Oryx. *Animal Kingdom* 86(Dec-Jan): 28-33.

The same information as in #188 but written for a popular and more accessible publication.

190. Flowerdew, J.R. 1987. *Mammals: their reproductive biology and population ecology*. London, Victoria, and Baltimore: Edward Arnold.

A basic book about mammals; their physiology, reproduction, behavioral factors affecting population dynamics, cyclical changes in numbers and population ecology. Discusses conservation issues in final three pages. Lists 26 rare mammal species identified as having self sustaining populations in zoos.

191. Flynn, R.W., and M.T. Abdullah. 1984. Distribution and status of the Sumatran rhinoceros in peninsular Malaysia. *Biological Conservation* 28(3): 253-273.

192. Fossey, D. 1983. *Gorillas in the mist: a remarkable woman's thirteen year adventure in a remote African rain forest with the greatest of the great apes*. London, Sydney: Hodder and Stoughton (Boston: Houghton Mifflin Co., 1984).

A very personal account with information about the behavior of the gorilla but more concerned with the author's life. Poachers are suspected of shooting her. This book was made into a movie.

193. Fossey, D. 1982. Reproduction among free-living mountain gorillas. *American Journal of Primatology*, Suppl. no. 1:97-104.

A scientific article that is less well-known to the public but which is the kind of work that led to Diane Fossey's reputation among scientists.

194. Fox, J.G., ed. 1988. *Biology and diseases of the ferret*. Philadelphia: Lea and Febiger.

The black-footed ferret is one of the most endangered mammals and is on the U.S. Endangered Species List. Book is a text and does not emphasize endangerment.

195. Franklin, K.E. 1987. Endangered species: where to from here? *American Forests* 93(11-12)Nov-Dec: 57-60, 74-77.

Claims the grizzly bear would have become extinct if it were not for the Endangered Species Act. Fine overview of the mechanics of the legislation, the need for greater funding, and the role of the U.S. Forest Service. Law has raised consciousness of the public and the government about biological diversity. See also **General** chapter item 33.

196. Friend, T. 1982. The numbat—an endangered specialist. *Australian Natural History* 20(10: 339-342.

An Australian on the U.S. Endangered Species List.

197. Froehlich, J.W., and P.H. Froehlich. 1987. The status of Panama's endemic howling monkeys. *Primate Conservation* 8:58-62.

198. Gautier-Hion, A., F. Bourliere, J.P. Gautier, and J. Kingdon, eds. 1988. *A primate radiation. Evolutionary biology of the African Guenons*. New York: Cambridge Univ. Press.

A collection of papers offering taxonomic interpretations and information about the ecology and behavior of old world monkeys.

* Genoways, H., and F.J. Brenner, eds. 1985. *Species of special concern in Pennsylvania.*

Cited in **General** Chapter as item 35.

199. Gosho, M.E., D.W. Rice, and J.M. Breiwick. 1984. The sperm whale, *Physeter macrocephalus. Marine Fisheries Review* 46(4): 54-64.

This issue is devoted to endangered whales.

200. Goudie, A. 1986. *The human impact on the natural environment.* 2nd ed. Cambridge, MA: MIT Press.

As part of this discussion of the environment, the human influence on the status of animals is explored both historically and in the present. Theories for causes of major periods of extinctions are presented, individual cases of species extinction are also shown as well as the effects of conservation policies.

201. Greene, J. 1985. Panthers at the vanishing point: Florida works to save America's most endangered mammals. *National Parks* 39(7-8)Jul-Aug: 18-24.

The count may be down to 30 panthers.

202. ______. 1983. Crossroads for the panther. *Defenders* 58(Sept-Oct): 6-11.

Author is not optimistic about saving the panther from extinction by automobile.

203. Groves, C.R., and T.W. Clark. 1986. Determining minimum population size for recovery of the black-footed ferret. *Great Basin Naturalist Memoir* 8:150-159.

* Groves, R.H., and W.D.L. Ride, eds. 1982. *Species at risk: Research in Australia.* Proceedings of a Symposium on the Biology of Rare and Endangered Species in Australia.

Cited in **General** Chapter as item 37.

204. Guiler, E. 1985. *Thylacine: The tragedy of the Tasmanian tiger.* Oxford: Oxford Univ. Press.

205. Gullan, P., and K. Norris. 1984. The habitat of *Burramys parvus* Broom in Victoria. In *Possums and Gliders,* edited by A. Smith and I. Hume, 417-421. Chipping Norton NSW, Australia: Surrey Beatty.

206. Hall, M.A., and S.D. Boyer. 1987. International Whaling Commission *Meeting Document* SC/39/SM12. Cambridge, U.K.: International Whaling Commission.

The IWC reports the killing of dolphins is at threatening levels.

207. Hallagan, J.B. 1981. Elephants and war in Zimbabwe. *Oryx* 16(2)Oct: 161-164.

Elephant population highest since 1940's, the only happy result of the war in Rhodesia which made the battle-torn bush dangerous for poachers. In addition, West Africa does not have an efficient system for ivory distribution as in East Africa.

208. Hancock, L. 1982. Vancouver Island marmot. *Nature* (Canada) 11(2): 18-23.

A Canadian mammal on the U.S. Endangered List.

209. Happold, D.C.D. 1987. *The mammals of Nigeria.* Oxford, U.K.: Clarendon Press.

Full descriptions of 247 species are provided in addition to the distribution and abundance of each species. Author cites need for conservation policy because of rapidly growing human population and increasing demand for land. Poaching is major concern.

210. Harcourt, A.H. 1981. Can Uganda's gorillas survive? A survey of the Bwindi Forest Reserve. *Biological Conservation* 19(4): 269-282.

Report of the most detailed study of its time on the numbers, distribution and conservation status of the gorilla in the Bwindi Forest Reserve, estimating a total population of 95 to 135 and a breeding population of 75.

211. Harris, S., ed. 1987. Mammal population studies. *Symposia of the Zoological Society of London* No. 58. Oxford: Clarendon Press.

212. Harrison, R., and M.M. Bryden, eds. 1988. *Whales, dolphins, and porpoises.* New York: Facts on File.

213. Heinsohn, G. 1986. World's only strictly marine sea cow threatened. *Australian Natural History* 21(2)Autumn: 530-531.

214. Heltne, P. 1985. *The Lion-tailed macaque: Status and conservation.* New York: A.R. Liss.

215. Henderson, J.R., and T.C. Johanos. 1988. Effects of tagging on weaned Hawaiian monk seal pups. *Wildlife Society Bulletin* 16(3): 312-317.

Population has declined 50% since 1958. Article is a report of a study to determine if tagging pups has a detrimental effect.

216. Hickling, G., and P. Hawkey. 1983. The grey seal, *Halichoerus grypus* (Fabr.) at the Farne Islands, 1981 and 1982. *Transactions of the Natural History Society of Northumbria,* No. 51.

217. Hicks, A. 1987. In search of wintering bats. *The Conservationist* (New York) 41(4): 14-17, 56-57.

Promotes the value of field surveys to confirm abundance or scarcity of endangered species, such as the Indiana bat.

218. Hoffmeister, D.F. 1986. *Mammals of Arizona.* [Tucson]: University of Arizona and Arizona Game and Fish Department.

219. Hood, C.S., L.W. Robbins, R.J. Baker, and H.S. Shellhammer. 1984. Chromosomal studies and evolutionary relationships of an endangered species, *Reithrodontomys raviventris*. *Journal of Mammalogy* 65(4): 655-667.

The use of molecular biology as a tool for studying the endangered salt marsh harvest mouse.

220. Horan, J. 1986. The red wolf is coming home. *Defenders* 61(3): 4-11.

221. Horwood, J.W. 1987. *The sei whale: population, biology, ecology, and management.* Croom Helm (UK); New York: Routledge Chapman & Hall.

Data on ecology, reproduction, biology, abundance and exploitation.

* Howard, R., A. Perry, P. Herring, and J. Scott (Study Manager). 1984. *Endangered and threatened species on U.S. Air Force installations.*

Cited in **General** Chapter as item 42.

222. Howe, J.R. 1983. The sea otter puzzle. *Audubon* 85(1)Jan: 34-39.

Listed as a threatened species and demonstrating declining numbers, the conflict over the future of California coastal waters between development interests and conservationist groups is described. The role of the sea otter in the ecology of the Aleutian Island chain demonstrates the complexity of conservation planning.

223. Humphrey, S.R., and D.B. Barbour. 1981. Status and habitat of three subspecies of *Peromyscus polionotus* in Florida. *Journal of Mammalogy* 62(4): 840-844.

Attention is paid to a Floridian mouse.

224. Humphrey, S.R., and B. Bell. 1986. The key deer population is declining. *Wildlife Society Bulletin* 14(3): 261-265.

225. Humphreys, W.F., R.A. How, A.J. Bradley, C.M. Kemper. and D.J. Kitchener. 1984. The biology of *Wyulda squamicaudata* Alexander 1919. In *Possums and Gliders,* eds., Smith, A. and I. Hume, 163-169. Chipping Norton NSW, Australia: Surrey Beatty & Sons.

The Scaly-tailed possum is on the U.S. Endangered List. Article describes its size, growth, development, distribution and abundance, and reproduction.

226. Inskipp, T., and J. Barzdo. 1987. *World checklist of threatened mammals.* Cambridge, UK: Conservation Monitoring Center, IUCN, and the Nature Conservancy Council.

Summary of basic information found in the *IUCN Red Data Books* or listed by CITES. Includes common names, country where species is found, and references to other sources.

* International Convention Advisory Commission. 1981. *Convention on international trade in endangered species of wild fauna and flora.* Washington, DC: CITES.

Cited in **General** chapter as item 43.

* IUCN Conservation Monitoring Centre. 1986. *1986 IUCN red list of threatened animals.*

Cited in **General** Chapter as item 44.

* IUCN Conservation Monitoring Centre. 1988. *1988 IUCN red list of threatened animals.*

Cited in **General** Chapter as item 45.

227. Jackson, P. 1987. Another tiger faces extinction. *Species* 8:17-19.

Argues that the Chinese tiger is endangered.

228. Jackson, P. 1986. Planning a future for the snow leopard. *WWF (World Wildlife Fund) Monthly Report* October: 279-283.

Report of an International Snow Leopard Trust meeting in Kashmir. Article describes habitats and range, behavior, threats to survival and a program for conservation adopted by group.

229. Jackson, P. 1983. The tiger lives on. *Ambio* 12(5): 278-279.

Description of the tiger's precarious status in India. Villagers fearful of its attacks.

230. Jameson, E.W., Jr., and H.J. Peeters. 1988. *California mammals.* California Natural History Guides, vol. 52. Berkeley: Univ. of California Press.

Not seen. Large number of mammals on both the federal and the California endangered lists.

231. Jeffrey, S. 1982. Threats to the proboscis monkey. *Oryx* 16(4)June: 337-339.

Describes their habitat as threatened by logging and settlement.

232. Johanos, T.C., A.K.H. Kam, and R.G. Forsyth. 1987. *Hawaiian monk seal on Laysan Island: 1984*. NOAA (National Oceanic and Atmospheric Administration) Technical Memorandum. Honolulu: National Marine Fisheries Service.

A 47-page technical memo reporting a study of the endangered Hawaiian monk seal centering on the census of adults and pups and mating behavior. Available from NTIS, PB87-180410/XAB.

233. Johnson, J.H., and A.A. Wolman. 1984. The humpback whale, *Megaptera novaeangliae*. *Marine Fisheries Review* 46(4): 30-37.

234. Jolly, A. 1988. Madagascar's lemurs: on the edge of survival. *National Geographic* 174(2)August: 132-161.

Comprehensive overview of a primate that is native to Madagascar only. Written in typical National Geographic style of personalizing individual lemurs and case studies. A good deal of information and spectacular photographs are included. Relates the extinction of the giant lemur and current government policy to protect the Aye-Aye and other lemurs.

235. Jones, M.L., S.L. Swartz, and S. Leatherwood, eds. 1984. *The gray whale, Eschrichtius robustus*. New York: Academic Press.

236. Joslin, P. 1984. Environmental limitations and future of the Asiatic lion. *Journal of the Bombay Natural History Society* 81(3)Dec: 648-664.

Listed by the IUCN in 1966, the lion is studied within the Gir Sanctuary to determine preservation policy. Description, natural history, and behavior are included.

237. Kappeler, M. 1984. The gibbon in Java. In *The lesser apes. Evolutionary and behavioural biology*. eds., H. Preuschoft, D.J. Chivers, W.Y. Brockelman, and N. Creel, 19-31. Edinburgh: Edinburgh Univ. Press.

One of several papers about the gibbon in this collection. Emphasis is on biology rather than endangerment.

238. Kasparek, M. 1986. On the historical distribution and present situation of gazelles, *Gazella* spp. in Turkey. *Zoology Middle East* 1:11-15.

Not seen. Difficult publication to find. Little information about species in Turkey in the English language.

* Kaufman, L., and K. Mallory. 1986. *The last extinction*. Cambridge, MA: New England Aquarium and MIT Press.

Cited in **General** chapter as item 50.

239. Kayes, R.J. 1985. *The decline of porpoises and dolphins in the southern North Sea: a current status report.* Greenpeace Research Report RR-14. Oxford: Political Ecology Research Group, Ltd., 34 Cowley Rd., Oxford 0X4 1H2.

Combines presentation of natural history and population status that is useful for naturalists and conservationists, with detailed chemical studies of interest to the scientific community.

240. Kenney, R.D. 1984. Distributional biology of the cetacean fauna of the northeast United States continental shelf (megaptera, humpback). *Dissertation Abstracts International* 45(11-b): 3394. Ph.D., University of Rhode Island.

Sightings collected by the Cetacean and Turtle Assessment Programs were studied to identify principal habitats for various species. The humpback's primary habitat is the western Gulf of Maine and its distribution was analyzed with density, temperature, season and other variables.

241. Kenyon, K.W. 1981. The Hawaiian monk seal: an endangered marine mammal. *Water Spectrum* 13(2): 1-6.

Not seen.

242. Kerle, J.A. 1984. Growth and development of *Burramys parvus* in captivity. In *Possums and Gliders,* ed. A. Smith and I. Hume, 409-412. Chipping Norton NSW, Australia: Surrey Beatty & Sons.

An opossum on the U.S. Endangered list.

243. Khan, M.A.R. 1984. Endangered mammals of Bangladesh. *Oryx* 18(3)July: 152-156.

Brief but specific survey of the status of mammals and the causes of their endangerment. Includes bibliography.

* Kiew, B.H. 1985. Malaysia's 10 most endangered animals, plants and areas. *Malayan Naturalist* 38(4)May: 2-3.

Cited in **General** chapter as item 51.

244. Kilpatrick, C.W., S.C. Forrest, and T.W. Clark. 1986. Estimating a genetic variation in the black-footed ferret—a first attempt. *Great Basin Naturalist Memoir* 8: 145-149.

245. King, J.F. 1987. Nature reserves of the USSR. *Sierra* 72(3)May/June: 38-45.

Information about endangerment and preservation activity in the Soviet Union has not been readily available before Glasnost. A survey of Soviet reserves arranged by land type. Animals protected by Soviet conservation policy discussed include walruses. Inventory of rare and endangered species published in 1974 and updated. Previous edition was 1985. The *Red Book (Krasnaya kniga)* follows IUCN classification system.

246. Kleiman, D.B., B.B. Beck, J.M. Dietz, L.A. Dietz, J.D. Ballou, and A.F. Coimbra-Filho. 1986. Conservation program for the golden lion tamarin: captive research and management, ecological studies, educational strategies, and reintroduction. In *Primates: the road to self-sustaining populations,* ed., K. Benirschke, 959-979. New York: Springer-Verlag.

247. Knight, R.R., and L.L. Eberhardt. 1984. Projected future abundance of the Yellowstone grizzly bear. *Journal of Wildlife Management* 48:1434-1438.

248. ______. 1985. Population dynamics of the Yellowstone grizzly bear. *Ecology* 66(2): 323-334.

A companion piece to the previous article.

249. Knight, R.R., B.M. Blanchard, and L.L. Eberhardt. 1988. Mortality patterns and population sinks for Yellowstone grizzly bears, 1973-1985. *Wildlife Society Bulletin* 16(2)Summer: 121-125.

A long-range study to identify causes of mortality of adult female grizzlies.

250. Kolstad, M., I. Mysterud, T. Kvam, O.J. Sorensen, and S. Wikan. 1986. Status of the brown bear in Norway: distribution and population in 1978-82. *Biological Conservation* 38(1): 79-99.

Report of a survey shows difficulty in counting bears. Estimated number resulting from survey is 160-230 individuals. Authors view bear as vulnerable but not threatened in Norway.

251. Kruuk, H. 1986. The case of the clannish badger. *Natural History* 95(12)Dec: 51-56.

Largest wild carnivore left in Britain, the relation between badgers and bovine disease has created conflict between conservation groups and farmers. Badger used as symbol by many conservation societies.

252. Kumar Mukherjee, A. 1982(1984). Spotlights on some endangered mammals of northeast India. *Indian Museum Bulletin* 17:24-40.

Not seen. Author is also cited as A.K. Mukherjee.

253. Ladd, W. 1986. New hope for the southern sea otter. *Endangered Species Technical Bulletin.* 11(8/9): 12-14.

254. Laidler, K., and L. Laidler. 1983. *The river wolf (Der Riesenotter).* London: George Allen & Unwin.

An account of the authors' efforts to save the giant otter (world's largest) from extinction.

255. Largen, M.J., and D.W. Yalden. 1987. The decline of elephant and black rhinoceros in Europe. *Oryx* 6(2): 103-106.

256. Lavery, H.J., and P.J. Tierney. 1985. Scarcity and extinction. In *The*

Kangaroo Keepers. Ed. H.J. Lavery, 25-52. St. Lucia, London & New York: Univ. of Queensland Press.

Case history of the bridled nailtail wallaby, a marsupial that had all but disappeared in Australia. Poses alternative approaches for conservation in Australia.

257. Lemke, T.O. 1986. Marianas fruit bats near extinction. *Bats* 3(1): 1-2

These bats are on the U.S. "endangered" list.

258. ______. 1986. Distribution and status of the sheath-tailed bat *Emballonura semicaudata* in the Mariana Islands. *Journal of Mammalogy* 67(4): 743-746.

Not on the U.S. "endangered" list but decline is noted.

259. Lewin, R. (C. Ross). 1988. Life history patterns emerge in primate study. *Science* 242(4885) Dec 23: 1634-1636.

Study of primates by C. Ross to explore reasons for differences in potential population increases among primates of varying sizes from the largest (gorilla) to the smallest (mouse lemur). Although article does not mention endangerment it does concern mammals known to be endangered.

260. Limpus, C.J., C.J. Parmenter, and C.H.S. Watts. 1983. *Melomys rubicola*, an endangered murid rodent endemic to the Great Barrier Reef of Queensland. *Australian Mammal* 6(2): 77-79.

261. Ljungblad, D.K., S.E. Moore, J.T. Clarke, and J.C. Bennett. 1987. *Distribution, abundance, behavior and biacoustics of endangered whales in the Alaskan Beaufort and Eastern Chukchi Seas*. Final Report. Aug-Oct 86. San Diego: Naval Ocean Systems Center, in cooperation with SEACO, Inc. and Minerals Management Service, Anchorage AK. Alaska Outer Continental Shelf Office.

Report of aerial surveys and acoustic recordings of bowhead whales, grey whales and other marine mammals to determine migration patterns, abundance and density, behavior, and sound production. Available from NTIS as PB88-116470/XAB.

262. Lloyd, P. 1984. The cape mountain zebra. *African Wildlife* 38(4): 144-149.

263. Luoma, J.R. 1987. The state of the tiger. *Audubon* 89(4)July: 61-63.

Names recently extinct tigers and those whose existence is precarious. Reports proceedings of a 1986 symposium sponsored by the Minnesota Zoological Gardens and the Species Survival Commission of the IUCN.

264. Macdonald, D. 1984. *Encyclopedia of mammals*. London and Sydney: George Allen and Unwin.

So many mammals are endangered, threatened, vulnerable or rare that any comprehensive collection will have relevant information.

265. Mace, G.M. 1988. The genetic and demographic status of the western lowland gorilla in captivity. *Journal of Zoology* (London) 216(4): 629-654.

Purpose of the study of gorillas in zoos is to explore the possibility of reproduction among existing captive gorillas and avoid capturing more from the wild to ensure perpetuation of species. Capture of young gorilla usually means adult parent is killed. High prices for young gorillas captured in the wild may stimulate killing of adults.

266. Macnamara, M.C. 1982. Huemul: a deer of distinction. *Animal Kingdom* 85(2)April-May: 6-13.

The huemul's gentle nature has made it an easy prey for human hunters in its native Chile. Author urges a protected reserve for its survival.

267. Maguire, L.A., T.W. Clark, R. Crete, J. Cada, C. Groves, M.L. Shaffer, and U.S. Seal. 1988. Black-footed ferret recovery in Montana: a decision analysis. *Wildlife Society Bulletin* 16(2)Summer: 111-120.

One of the most critically endangered species in the world. Thought to have become extinct in 1970, it was rediscovered in Meeteetse, Wyoming. A recovery program of captive breeding was begun but dissatisfaction led authors to develop a model for making decisions under uncertainty, balancing concern for protection of habitat and captive breeding.

* Majumdar, S.K., F.J. Brenner, and A.F. Rhodes, eds. 1986. *Endangered and threatened species programs in Pennsylvania and other states: causes, issues, and management.* Harrisburg: Pennsylvania Academy of Science.

Cited in **General** chapter as item 54.

268. Mallinson, J. 1984. Lion Tamarins' survival hangs in balance. *Oryx* 18(2)April: 72-79.

269. ______. 1986. Plight of a pigmy hog. *Country Life* 179(4635): 1762-1763.

Not seen.

270. ______. 1986. A note on the successful breeding and reintroduction of *Leontopithecus rosalia* with special reference to an international recovery programme and management committee for *Leontopithecus chrysamelas. Primate Eye* no. 29 (Supple): 48-52.

Describes a successful project for tamarins.

271. Malpas, R. 1981. Elephant losses in Uganda and some gains. *Oryx* 16(1)May: 41-44.

Effects of political situation on status of elephants: warfare made poaching inactive.

272. Mansergh, I. 1984. The mountain pygmy possum *Burramys parvus* Broom: a review. In *Possums and Gliders.* ed. A. Smith and I. Hume, 413-416.

Chipping Norton NSW, Australia: Surrey Beatty and Sons.

273. Mansergh, I. 1984. Ecological studies and conservation of *Burramys parvus*. In *Possums and Gliders* eds. Smith, A., and I. Hume, 545-552. Chipping Norton NSW, Australia: Surrey Beatty and Sons.

A marsupial on the U.S. Endangered List.

274. Marine Mammal Commission. 1987. *Annual report of the Marine Mammal Commission, calendar year 1986*. Washington, D.C.: Marine Mammal Commission.

The Marine Mammal Commission is required to present an annual report to the Congress under the Marine Mammal Protection Act of 1982. The report has detailed information about the current status of whales, seals, dolphins and other marine mammals.

275. Martin, E.B. 1984. They're killing off the rhino. *National Geographic Magazine* 165(3)March: 404-422.

Report on why and how the rhino is being exterminated accompanied by photographs of its habitat and behavior.

276. Martin, E.B., and C.B. Martin. 1982. *Run rhino run*. London: Chatto & Windus.

The rhino in history and mythology, the antisocial rhino, the slaughter of rhinos (Asian and African), the use of rhino products in business, the rhino in African culture and current conservation efforts, and illustrations. What more could one ask.

277. Martin, M. 1983. The California sea otter: endangered, threatened, or thriving? *Environment* (St. Louis) 25(7)Sept: 28-32.

Report of a 5-year review coming up to assess appropriateness of "threatened status" on the ESL. Supporters of "threatened" status point to danger of oil spills from the off-shore oil tanker traffic and the presence of 2 tanker ports within the otter's range. Proposals to move the otter rejected by shell fisherman near San Nicholas as the otter threatens their catch.

278. Mason, C.F., and S.M. Macdonald. 1986. *Otters: Ecology and conservation*. New York: Cambridge Univ. Press.

Sharp decline in otter population of Britain gave rise to interest in the 13 known species. Authors describe animal's current status, biology, and cause for decline. Otters are dependent on river habitat and their abundance is an indicator of the health of the river.

279. McCracken, G.F. 1987. Why are we losing our Mexican free-tailed bats? *Endangered Species Technical Bulletin Reprint* 4(4): 1-3. Univ. of Michigan, School of Natural Resources.

Pesticides, the loss of caves from human disturbance, and epidemic disease are considered causes for decline. Article describes reproduction, diet, and migratory patterns.

280. McDonald, S., and N. Duplaix. 1983. The otter, symbol of our threatened fauna. *Naturopa* 45:14-19.

There are a considerable number of publications about the European otter and this one is a good summary of the otter's diet, habitat needs, European distribution and causes of its decline. The journal is a publication of the European Information Centre for Nature Conservation.

281. McNamee, T. 1982. Breath-holding in grizzly country. *Audubon* 84(6)Nov: 69-83.

Anecdotal account of traveling through Alaskan wilderness presented in a portrait of grizzlies intended to convince of the need for conservation.

282. ______. 1984. *The Grizzly bear.* New York: Alfred A. Knopf, Inc.

Complete portrait of the grizzly plus an exhaustive bibliography. Reports that Yellowstone National Park has 200 bears but the number is declining.

283. McNeely, J.A., and P.S. Wachtel. 1988. *Soul of the tiger.* New York: Doubleday.

Exploring southeast Asian civilization and the approach to ecology, the authors describe the position of tiger as predator, as danger to villagers, and having market value dead or alive. These roles are in conflict with preservationist.

284. Mech. L.D. 1988. *The Arctic wolf: living with the pack.* Stillwater, MN: Voyageur Press.

The author has spent 30 years studying this endangered species and poses ethical issues about its existence. In addition to the natural history and biology of the Arctic wolf, the author proposes international cooperation as the means to preserve this and other species. Well illustrated.

285. Meith, N. 1984. Saving the small cetaceans. *Ambio* 13(1):2-14.

Argues that large whale species used for commercial purposes will be helped by the moratorium treaty but the related, smaller porpoises, dolphins and small whales are threatened due to scientific ignorance. Lists include scientific and common names, and categories of risk: endangered, vulnerable, rare, out-of-danger, and indeterminates. Provides a historical overview of causes of decline, protective measures, international meetings and conventions since 1940. Contains good bibliography.

286. Merritt, J.F. 1987. *Guide to Mammals of Pennsylvania.* Pittsburgh: Univ. of Pittsburgh Press.

Species' descriptions include the classifications of vulnerable, extirpated, and undetermined.

288. Miller, S.D., J. Rottmann, K.J. Raedeke, and R.D. Taber. 1983. Endangered mammals of Chile: status and conservation. *Biological Conservation* 25(4) April: 335-352.

Account of 78 species with a designation of their status in the terms used by the IUCN. Of the 78 native mammals, 22 species plus 5 subspecies are of conservation concern. Major problem is lack of enforcement of protection laws.

289. Mills, S. 1987. Does Sweden have room for its wolves? *Oryx* 6(2): 92-96.

Description of habitat needed for wolf behavior.

290. Minasian, S.M., K.C. Balcomb, and L. Foster. 1984 . *The world's whales: the complete illustrated guide*. Washington, D.C.: Smithsonian Institution Press. Distributed by Norton.

291. Mittermeier, R.A. 1987. Current distribution of the muriqui in the Atlantic Forest region in eastern Brazil. *Primate Conservation* 8:143-149.

292. ______. 1981-82. A shaggy-monkey story. *Animal Kingdom* 84(6)Dec-Jan: 26-31.

Author considers the white monkey Uakari endangered in the Amazon forest.

* ______. 1981. Brunei protects its wildlife. *Oryx* 16(1)May: 67-70.

Effective laws are preserving the proboscis monkey, clouded leopard, and the dugong. Also cited in **General** Chapter as item 57.

294. Mizroch, S.A., D.W. Rice, and J.M. Breiwick. 1984. The blue whale, *Balaenoptera musculus*/ The fin whale, *Balaenoptera physalus*/ The sei whale, *Balaenoptera borealis*. *Marine Fisheries Review* 46(4): 15-19, 20-24, 25-29.

Entire issue of this periodical is concerned with endangered whales.

* Mowat, F. 1984. *Sea of slaughter*. Boston: Atlantic Monthly. New York: Bantam Books, repr. 1986.

Buffalo, curlews, and whales among the several mammals described. Cited in **General** Chapter as item 58.

* Mukherjee, A.K. 1982. *Endangered animals of India*. Calcutta: Zoological Survey of India.

Cited in **General** Chapter as item 59.

295. Munn, M.B., and C.A. Munn. 1988. The Amazon's gregarious giant otter. *Animal Kingdom* 91(5)Sept-Oct: 34-41.

Article describes the group living of the South American otter, a behavior different from the North American and European otters.

296. Murray, J.A. 1987. *Wildlife in peril: the endangered mammals of Colorado.* Boulder: Roberts Rinehart.

A good deal of natural history of many mammals and the history of their decline.

297. Nawaz, M. 1983. The endangered mammals of Pakistan. *Tigerpaper* 10(3): 15-20.

298. Ness, P.S. 1986. Introducing the West Indian manatee. *Freshwater and Marine Aquaria* 9(10): 50-51.

299. Nielsen, C.E. 1985. Occurrence of *Lutra lutra* in the north of Jutland in the autumn of 1984. [In Danish with English summary.] *Flora Fauna, Silkeborg* 91(3-4): 13-15.

* Nilsson, G. 1983. *The endangered species handbook.* Washington, D.C.: Animal Welfare Institute.

See item 63 in **General** Chapter.

* Norden, A.W., D.C. Forester, and G.H. Fenwick. 1984. *Threatened and endangered plants and animals of Maryland.* Special Publication/ Maryland Natural Heritage Program: 84-I. Annapolis, MD: Department of Natural Resources.

Cited in **General** Chapter as item 64.

* Norris, R. 1988. Data for diversity. *Nature Conservancy Magazine* 38(1)Jan-Feb: 4-10.

The Brazilian three-toed sloth is among the animals mentioned. Cited in **General** Chapter as item 65.

300. Nysewander, D.R., D.J. Forsell, P.A. Baird, D.J. Shields, G.W. Weiler, and J.H. Kogan. 1982. *Marine bird and mammal survey of the eastern Aleutian Islands, summers of 1980-1981.* Unpublished report. Anchorage, Alaska: U.S. Fish and Wildlife Service.

A base-line study that can be used as a comparison with previous studies and those of the future. Not seen. Should be available by writing to the department.

301. O'Brien, S.J. 1987. The ancestry of the giant panda. *Scientific American* 257(5)Nov: 102-107.

Story of a taxonomic controversy. Author believes the giant panda is a bear in the Ursidae family, while the red panda, he argues, belongs with the raccoons or the Procyonidae family.

302. O'Brien, S.J. et al. 1983. The cheetah is depauperate in genetic variation. *Science* 221(4609) July 29: 459-462.

Article examines biochemical genetic variation in samples of two geographically isolated South African cheetah populations.

303. O'Brien, S.J., D.E. Wildt, and M. Bush. 1986. The cheetah in genetic peril. *Scientific American* 254(5)May: 84-92.

Report of a genetic analysis of the cheetah that posits a genetic bottleneck in the Pleistocene that ended genetic diversity. It may have eliminated the most deleterious genes and authors propose breeding in captivity with subspecies.

304. O'Brien, S.J., M.E. Roelke, L. Marker, A. Newman, C.A. Winkler, D. Meltzer, L. Colly, J.F. Evermann, M. Bush, and D.E. Wildt. 1985. Genetic basis for species vulnerability in the Cheetah. *Science* 227(4693)March 22: 1428-1434.

A discussion of the circumstances leading to the decline of this animal, the difficulties in artificial propagation techniques, and the possible hypersensitivy to the cheetah of a viral pathogen, coronavirus.

305. O'Shea, T.J., C.A. Beck, R.K. Bonde, H.I. Kochman, and D.K. Odell. 1985. An analysis of manatee mortality patterns in Florida, 1976-81. *Journal of Wildlife Management* 49(1): 1-11.

306. Oates, J.R. 1982. In search of rare forest primates in Nigeria. *Oryx* 16(5)Oct: 431-436.

Urges listing of the Red-bellied guenon in the *Red Data Book* and Appendix I of CITES.

307. Oliver, W.L.R. 1984. Distribution and status of the hispid hare. *Dodo* 21:6-32.

On the U.S. Endangered List.

308. Osborne, B.C., D.P. Mallon, and S.K.R. Fraser. 1983. Ladakh: threatened stronghold of rare Himalayan mammals. *Oryx* 17(4)Oct: 182-189.

Primarily an article identifying animals that have become rare in Indian Ladakh: their diet and habitat, and current distribution. References are included.

309. Ottichilo, W.K. 1987. The causes of the recent heavy elephant mortality in the Tsavo ecosystem, Kenya, 1975-1980. *Biological Conservation* 41(4): 279-289.

Any information that contributes to our knowledge of the decline of these unique and popular animals is useful.

310. Parfit, M. 1985. Its days as a varmint are over but the cougar is still on the run. *Smithsonian* 16(6)Sept: 68-78.

An account of scientists studying the cougar, also called the mountain lion. It is the Americas' second largest cat only slightly smaller than the jaguar. Lions are protected in the Guadalupe Mountains National Park in Texas but can be killed outside the park. Article reviews some natural history of the cougar, conflicts with ranchers, and the varying state laws regarding hunting.

311. Parker, I.S.C., and E.B. Martin. 1982. How many elephants are killed for the ivory trade? *Oryx* 16(3)Feb: 235-239.

Trade appears to have declined between 1976 and 1980 but statistics based on import data in Hong Kong and Japan rather than export data from Africa.

312. Peek, J. et al. 1987. Grizzly bear conservation and management: a review. *Wildlife Society Bulletin* 15(2): 160-169.

313. Penny, M. 1987, 1988. *Rhinos: endangered species.* London: Christopher Helm; New York: Facts on File.

Current status of five rhino species. A biological description in the form of a literature survey. Half the book devoted to causes of decline (conflict with man) and conservation efforts.

314. Perrin, W.F. 1988. *Dolphins, porpoises and whales: an action plan for the conservation of biological diversity, 1988-1992.* Gland, Switzerland: IUCN.

315. Peyton, B. 1981. Spectacled bears in Peru. *Oryx* 16(1)May: 48-56.

Listed as vulnerable in the IUCN Red Data Book, the bear inhabits the Andean Range from western Venezuela to Bolivia; isolated areas in Ecuador, southern Colombia and Peru. A combination of hunting and cutting down of scrub desert and thorn forest contributed to decline. Concern that isolated populations will lack genetic diversity for continued reproduction.

316. Phillips, M.K., and W.T. Parker. 1988. Red wolf recovery: a progress report. *Conservation Biology* 2(2)June: 139-141.

A successful captive breeding program produced 80 wolves. Authors describe a 5-year plan that was begun in 1987 to establish them into the wild in North Carolina taking behavior and diet into consideration.

317. Pilleri, G., and O. Pilleri. 1981. Some observations on mammals, birds and reptiles in the lower valley of the Indus and upper valley of the Brahmaputra. *Bollettino Soc. Adriat. Sci. Nat.* 65(2): 133-166.

Not seen but cited in *Zoological Record.*

318. Pippard, L. 1985. Status of the St. Lawrence River population of Beluga, *Delphinapterus leucas. Canadian Field Naturalist* 99(3): 438-450.

The St. Lawrence River population is endangered but beluga are abundant in the Canadian Arctic. Over-hunting in the 16th century was the initial cause of the decline (hunting ceased in 1975) but loss of critical habitats in certain estuaries, isolated populations, environmental contaminants and other factors are described in detail.

319. Poole, J.H. 1987. Elephants in musth, lust. *Natural History* 96(11)Nov: 46-55.

Elephant sexual behavior.

320. Povilitis, A. 1983. The huemul in Chile: national symbol in jeopardy. *Oryx* 17(1)Jan: 34-40.

The huemul is a cultural symbol in Chile but its range and habitat are being restricted. The author provides a good physical description, conservation requirements including diet, current range and distribution, and photographs.

321. Povilitis, J. 1986. Grizzly bear mortality in Yellowstone: implications for conservation. *Western Wildlands* 12(4): 15-18.

322. Preuschoft, H., D.J. Chivers, W.Y. Brockelman, and N. Creel., eds. 1984. *The lesser apes: evolutionary and behavioural biology*. Edinburgh: Edinburgh Univ. Press.

A collection of scientific papers from a conference about the lesser apes: i.e., chimpanzees, gibbons, and orangutans. Articles in this volume are mostly about gibbons in the wild and in zoos. Ecology, social behavior and evolutionary biology are frequent topics. An extensive bibliography.

323. Pryde, P.R. 1986. Strategies and problems of wildlife preservation in the USSR. *Biological Conservation* 36(4): 351-374.

An overview of the USSR *Red Book* (1974, rev 1978-1985) which lists 25 mammals as endangered. They face the same issues found elsewhere: loss of habitat and pollution.

324. Rabinowitz, A. 1986. *Jaguar: struggle and triumph in the jungles of Belize*. New York: Arbor House.

As much an account of Belize as it is of jaguars. Readable personal account of searching for jaguars for the purpose of establishing a preserve. The successful result was the establishment of the Cockscomb Basin as a National Forest Reserve.

325. ______, and B. Nottingham. 1986. Ecology and behavior of the jaguar, *Panthera onca*, in Belize, Central America. *Journal of Zoology* (London) 210(a): 149-159.

326. Ralls, K., J.D. Ballow, and A. Templeton. 1988. Estimates of lethal equivalents and the cost of inbreeding in mammals. *Conservation Biology* 2(2)June: 185-193.

Captive breeding has shown that inbreeding reduces the survival and the fecundity of the animals but, theoretically, it ought to be possible to produce the most fit when the gene structure is known. Experiments to determine when inbreeding has the greatest costs, as defined by lack of survival and fecundity, showed too wide a variety between individuals to arrive at a specific conclusion.

327. Randall, D. 1986. Survival crisis at Meeteetse. *Defenders* 61(1) Jan-Feb: 4-11.

The story of the discovery of the ferret colony in Wyoming laid to rest concerns for ferret survial. However, sylvatic plague spread by fleas to prairie dogs drastically cut down ferrets' food supply. Evaluation of management policy concludes that six ferrets should have been taken from the colony for captive breeding.

328. Rathbun, G.B., C.A. Woods, and J.A. Ottenwalder. 1985. The manatee in Haiti. *Oryx* 19(4): 234-236.

329. Rathbun, G.B., J.A. Powell, and G. Cruz. 1983. Status of the West Indian manatee in Honduras. *Biological Conservation* 26(4): 301-308.

Animal is considered endangered in southeastern U.S., Caribbean Sea and South America.

330. Reeves, R. 1982. What hope for the North Atlantic right whales. *Oryx* 16(3)Feb: 255-262.

The right whale, once believed to be extinct, has been "rediscovered" several times. The tranquil area of the Bay of Fundy provides the environment for a secure recovery.

331. Reiger, G. 1987. Too many cooks. *Field and Stream* 92(7)Nov: 15-16.

Criticizes federal scientific wildlife management for being too manipulative. Favors setting aside undisturbed habitat reserves. Cites case histories of grizzlies, black-footed ferret, Florida panther and California condor.

332. Reynolds, J.E., III, and J.R. Wilcox. 1987. People, power plants and manatees. *Sea Frontiers* 33(JulAug): 263-269.

Not seen.

333. Ricciuti, E.R. 1982. The vicuna—victor or victim? A special investigation controversy. *Animal Kingdom* 85(3)June-July: 6-23.

The value of its pelts brought the vicuna to the brink of extermination but their trade has been banned by CITES. Diet and grazing areas in competition with sheep, the campesino's livelihood. Good natural history profile included.

334. Rice, D.W., A.A. Wolman, and H.W. Braham. 1984. The gray whale, *Eschrichtius robustus*. *Marine Fisheries Review* 46(4): 7-14.

335. Richardson, L., T.W. Clark, S.C. Forrest, T.M. Campbell, III. 1986. Black-footed ferret recovery: a discussion of some options and considerations. *Great Basin Nature Memoir* 8:169-184.

336. Richardson, R. 1986. Decline and fall of the blackfooted ferret. *Natural History* 95(2)Feb: 69-77.

On the track of the last of the black-footed ferrets. Author describes the link with the prairie dog, the ferret's prey. This periodical is found in almost every library.

337. Ride, W.D.L., and G.R. Wilson. 1982. *Australian animals at risk.* In *Species at risk-research in Australia.* Appendix 1. edited by R.H. Groves and W.D.L. Ride. New York: Springer-Verlag.

Volume cited in **General** chapter as item 37.

338. Robbins, J. 1988. When species collide: grizzly and man. *National Wildlife* 26(2)Feb-Mar: 21-27.

Controversy between those who see the grizzly as a threat to man and urge its "delisting" as an endangered species, and those who wish to continue protection of the bear. The latter group cites the need for understanding the bear's behavior.

339. _______. 1986. Wolves across the border. *Natural History* 95(5)May: 6,8,10,12-15.

Raises the issue of whether predators should be protected.

340. Roberts, L. 1988. Conservationists in Panda-monium. *Science* 241(4865)Jul 29: 529-531.

World Wildlife Fund sues U.S. Fish and Wildlife Service for allowing Toledo, Ohio Zoo to borrow breeding pandas for exhibit. Controversy points up conflicting approaches to preventing extinction of the panda. Article describes status of panda in small, isolated wildlife preserves that prevent genetic diversity and minimal fertility in zoo or captive areas.

341. _______. 1988. Beyond Noah's Ark: what do we need to know? *Science* 242(4883)Dec 2: 1247.

A single page news item reporting on a workshop sponsored by the National Science Foundation to "examine research priorities in single species conservation biology." In so doing, the report identifies some major issues such as conditions required for adaptation to the wild, human cooperation in preserving habitats, viable small population size, genetics and demography, role of stress and reproductive biology.

* Roots, C. 1987. *Endangered species: Canada's disappearing wildlife.* Illus., C. Godkin. Markhan, Ont.: Fitzhenry & Whiteside.

Primarily an identifying list of 101 lost species or those that may disappear soon; their description and causes of endangerment. Item 75 in **General** chapter.

342. Rosenberg, S. 1987. Sea otter vs. fisherman. *Skin Diver* 36(Feb): 140-147.

Author describes this "hotly contested California controversy" which centers on the otter's diet.

343. Ryder, O.A., P.C. Brisbin, A.T. Bowling, and E.A. Wedemeyer. 1981. Monitoring genetic variation in endangered species. In *Evolution today. Proceedings of the Second International Congress of Systematic and Evolutionary Biology,* University of British Columbia, Vancouver, July 1980. Edited by G.G.E. Scudder and J.L. Reveal, 417-424. Pittsburgh: Carnegie-Mellon University, Hunt Institute for Botanical Documentation.

The progress of evolution in species whose existence is dependent on captive breeding is of current interest.

344. Rylands, A., and R.A. Mittermeier. 1983. Parks, reserves, and primate conservation in Brazilian Amazonia. *Oryx* 17(2)Apr: 78-87.

Lists animals, parks, reserves and their vegetation types. Article is important because of the role of development in this critical area.

* Saharia, V.B. 1982. *Wildlife in India*. Dehra Dun, India: Natraj Publishers.

Cited in **General** Chapter as item 76.

345. Sandhu, J.S., and M.S. Dhindsa. 1986. On inbreeding and conservation of white tiger. *Tigerpaper* 13(4): 25-27.

346. Santiapillai, C., and W. Sukohado Ramono. 1985. On the status of the tiger (*Panthera tigris sumatrae* Pocock, 1829) in Sumatra. *Tigerpaper* 12(4): 23-29.

Not seen. Cited in *Zoological Record.*

347. ______, and K. Asby. 1988. The clouded leopard in Sumatra. *Oryx* 22:44-45.

Summary of current status.

348. ______, and H. Suprahman. 1986. The proposed translocation of the Javan Rhinoceros, *Rhinoceros sondaicus. Biological Conservation* 38(1): 11-19.

Only 54 Javan rhinoceros survive in a reserve in Java and exist nowhere else, not even in zoos. Controversial plan to transfer some Javan rhinos to a reserve in Sumatra is considered.

* Sarkar, S.U. 1985. Endangered wildlife of the coastal zone of Bangladesh. *Coastal Zone* 1985(1): 1110-1120.

Cited in the **General** Chapter as item 78.

349. Schaaf, C.D., and M.D. Stuart. 1983. Reproduction of the mongoose lemur (*Lemur mongoz*) in captivity. *Zoo Biology* 2(1): 23-38.

An example of the activity to preserve species through breeding in zoos.

350. Schaller, G.B. 1986. Secrets of the wild panda. *National Geographic Magazine* 169(3)March: 284-309.

Written in a popular style that reports the results of putting radio monitors on pandas in 1981. In relaying observations about the study and two pandas in particular, much information is passed on regarding the feeding habits, reproduction, and social behavior—all accompanied by the usual fine National Geographic photographs.

351. Schaller, G.B., H. Jinchu, P. Wenshi, and Z. Jing. 1985. *Giant pandas of Wolong*. Chicago: Univ. of Chicago Press.

Slightly over 1000 pandas are left in the wild. Authors state that loss of bamboo habitats and the isolation of panda groups contribute to the decline. Attempts by zoos to breed pandas are newsworthy items but cubs are rare and mortality is high. Authors suggest that the long gestation period, the single birth, and the dependent cub are the result of a history of plentiful food and space that "obviated the need to invest energy in reproduction."

352. Schwartz, J.H., ed. 1988. *Orangutan biology*. New York: Oxford Univ. Press.

Information about the behavior, ecology, reproductive biology and anatomy are some aspects of orangutan biology discussed. The primary emphasis is, however, on the phylogenetic affiliation between the orangutan and humans.

353. Servheen, C. 1985. The grizzly bear. In *Audubon wildlife report 1985* ed. by R. DiSilvestro, 400-415. New York: National Audubon Society.

354. Shaffer, M.L., and F.B. Samson. 1985. Population size and extinction: a note on determining critical population sizes. *American Naturalist* 125(1)Jan: 144-151.

Use of the Yellowstone grizzly as an example of their theory in developing a model for determining minimum viable population size. Cited in **General** Chapter as item 81.

355. Shapiro, W.A. 1984. Current status and management of gray bat caves in Jackson County, Florida. *Florida Field Naturalist* 12(1): 1-6.

Information about the behavior and diet of this bat listed on the U.S. Endangered Species List.

355a. Sharma, R.K. 1985. Results of the 1981 survey for gharial (*Gavialis gangeticus*) in North National Chambal Sanctuary. *Journal of the Bombay Natural History Society* 82(1): 215-218.

356. Sharp, I. 1985. The rhino export deal. *Malayan Naturalist*. 39(1/2)Dec: 11-14.

Key issues involved in the decision to send an endangered Sumatran rhinoceros to American zoos for breeding in captivity. Final agreement is included.

357. Shaul, W., and A. Haynes. 1986. Manatees and their struggle for survival. *Jamaica Journal* 19(3): 29-36.

358. Silber, G.K. 1987. Can the vaquita be saved? *Defenders* 62(3)May-June: 14–15.

Article about a little-known marine mammal, the Gulf of California harbor porpoise, that the author claims was listed as endangered in 1985. He states that the use of gillnets at a totoaba (also endangered) fishery is causing the decline.

359. _______. 1988. Recent sightings of the Gulf of California harbor porpoise, *Phocoena sinus. Journal of Mammalogy* 69(2): 430-433.

* Skoog, P.J. 1982. *Highways and endangered wildlife in Florida: a manual of information and recommendations.* Tallahassee: Florida Game and Fresh Water Fish Commission and Florida DOT.

Death by automobile is contributing to extinction of Florida panther. Author suggests highway design that will keep panthers away from the highways. Also cited in **General** Chapter as item 82.

360. Smit, C.J., and A. Van Wijngaarden. 1981. *Threatened mammals in Europe.* [In English] Supplementary volume of *Handbuch der Saugetiere Europas.* [In German] Wiesbaden: Akademische Verlagsgesellschaft.

361. Smith, A., and I. Hume, eds. 1984. *Possums and gliders.* Prepared from the Australian Mammal Society's Possum and Glider Symposium held in Armidale, Australia, 1983. Chipping Norton, NSW, Australia: Surrey Beatty & Sons.

A collection of 57 research and review papers describing the life history, taxonomy, physiology, behavior and conservation of Australian forest dwelling marsupials. Includes several papers on the scaly-tailed possum and the mountain pygmy possum.

* Steadman, D.W., and S. Zousmer. 1988. *Galapagos: discovery on Darwin's island.* Washington, D.C.: Smithsonian Institution Press.

See item 85 of **General** Chapter.

* Stermer, D. 1981. *Vanishing creatures, a series of portraits.* With introduction by Robert Redford. Berkeley, Calif.: Lancaster-Miller Publishers.

See **General** Chapter, item 86.

362. Stone, G.S., S.D. Kraus, J.H. Prescott, and K.W. Hazard. 1988. Significant aggregations of the endangered right whale, *Eubalaena glacialis,* on the continental shelf of Nova Scotia, Canada. *Canadian Field Naturalist* 102(3): 471-474.

363. Stuart, C.T. 1985. The status of 2 endangered carnivores occurring in the

Cape-province, South Africa: felis-serval and lutra-maculicollis. *Biological Conservation* 32(4): 375-382.

As the result of a survey, the Felis-serval is now thought be be extinct. It may be possible to save the lutra if the river quality is improved.

364. Susman, R.L., N. Badrian, A. Badrian, and N.T. Handler. 1981. Pygmy chimpanzees in peril. *Oryx* 16(2)Oct: 179-183.

Threats to the existence of this species are considered so severe by the World Wildlife Fund it has made this animal a high priority in its Tropical Forest and Primates campaign.

365. Sussman, R., and R. Alison. 1986. Lemur conservation in Madagascar: the status of lemurs in the south. *Primate Conservation* 7:86-91.

366. Tattersall, I. 1982. *The primates of Madagascar.* New York: Columbia Univ. Press.

Lemurs are among the endangered primates described.

367. Telander, R. 1981. Riding herd on Peru's vicunas. *International Wildlife* 11(3): 36-43.

368. Tenaza, R. 1987. The status of primates and their habitats in the Pagai Islands, Indonesia. *Primate Conservation* 8:104-110.

369. Thomson, C.E. 1982. *Myotis sodalis. Mammalian Species* 163:1-5.

The Indiana Bat is on the U.S. Endangered List.

370. Thornback, J., and M. Jenkins, Compilers. 1982. *Threatened mammalian taxa of the Americas and the Australasian zoogeographic region (excluding cetacea). The IUCN Mammal Red Data Book,* Part 1. With the help and advice of the Species Survival Commission of IUCN and other experts. Gland, Switzerland: IUCN and Cambridge, U.K.: IUCN Conservation Monitoring Center.

The series has become the authoritative international register of threatened species. Reliable data on each species.

371. Thorne, E.T. 1986. Captive breeding and the single ferret. *Wyoming Wildlife* 50(12): 29-35.

Not seen.

372. Thorne, E.T., and E. Williams. 1988. Disease and endangered species: the black-footed ferret as a recent example. *Conservation Biology* 2(1)March: 66-74.

Decline usually attributed to agriculture replacing prairies and systematic eradication of prairie dogs. Restoration program pointed to role of disease, especially canine distemper, in death of captive breeding population of ferrets and the effect of disease on other rare species. Fine bibliography.

* Tikader, B.K. With a foreword by T.N. Koshoo. 1983. *Threatened animals of India*. Calcutta: Zoological Survey of India.

Tigers, lions, deers and hispid hares are included. Cited in **General** chapter as item 89.

373. Tilson, R.L., and U.S. Seal, eds. 1987. *Tigers of the world: the biology, biopolitics, management and conservation of an endangered species*. Proceedings of a symposium held at the Minnesota Zoological Garden and co-sponsored with the IUCN in 1986. Park Ridge, N.J.: Noyes Publications.

Of 8 subspecies, 3 have succumbed to extinction during this century. The China tiger and the Amur tiger may have fragmented populations too small to maintain the genetic diversity necessary for long term survival. 40 authorities contributed papers to identify the management, conservation, and research needs of this predatory animal.

374. Tixier, S. 1988. What place for the grizzly. *Endangered Species Update* 5(12): 1-4.

Not seen.

375. Torres, H. 1986. *The Guanaco: distribution and conservation*. South American Camelid Specialist Group, SSC, IUCN Special Report No. 2. Cambridge, U.K.: IUCN.

Not seen.

376. Turbak, G. 1984. Grizzly on the ropes. *American Forests* 90(2)Feb: 22-23,48-50.

Distribution of the grizzlies in North America is described.

377. ______. 1982. The cougar's new cloak. *National Wildlife* 20(3)Apr-May: 47-54.

Author gives a favorable view of the mountain lion's role in the environment. Makes light of its predatory activities on deer and farm animals. Same animal known by the four names: cougar, puma, panther, mountain lion.

378. U.S. Congress. Senate. Committee on Commerce, Science, and Transportation. *Marine Mammal Protection Act Reauthorization Hearings*. 98th Congress, 2nd sess., 1984. H.R. 98-785.

Includes information about the status of marine mammals and their need for protection as well as being an example of a source of information about endangered species in federal documents.

* U.S. Fish and Wildlife Service. 1987. *Restoring America's wildlife*. H. Kallman, Chief Editor. Washington, D.C.: Government Printing Office.

Cited in **General** chapter as item 49.

379. U.S. Department of the Interior, Bureau of Land Management. 1982 (June). *Twin Falls livestock grazing, Twin Falls County,Idaho*. Environmental Impact Statement. 2 vol. Burley, Idaho: Department of the Interior, Bureau of Land Management.

An environmental impact statement that analyzes a plan to increase grazing areas on public land. Discusses the effects of encroachment into the habitats of ground-nesting birds, winter and summer ranges for mule deer.

380. U.S. Department of the Army (Corps of Engineers) and Fish and Wildlife Service. 1982. *Bottomland hardwood preservation program, Reeves Bottom (Hornersville Swamp), Dunklin County, Missouri*. Jackson, Miss.: U.S. Department of the Interior, Fish and Wildlife Service.

Proposal to acquire a hardwood forest to be managed as a wildlife conservation area that would affect a habitat for several endangered species. Provides a useful description of the needs of those birds and mammals. Publication was prepared as an environmental impact statement.

381. Van Aarde, R.J., and J.D. Skinner. 1986. Pattern of space use by relocated servals *Felis serval*. *African Journal of Ecology* 24(2): 97-101.

Not seen.

382. Van Blaricom, G.R., and R.J. Jameson. 1982. Lumber spill in central California waters: implications for oil spills and sea otters. *Science* 215(4539): 1503-1505.

383. ______, and J.A. Estes, eds. 1988. *Community ecology of sea otters. Ecological Studies*, Vol. 65. New York: Springer-Verlag.

Listed as "threatened" by the U.S., its preservation from near extinction is a success story. This book considers the sea otter in its ecosystem as it may have "keystone" status, i.e., greatly influencing its community. The sea otter eats sea urchins who eat kelp, etc.

384. Vandenbeld, J. 1988. *Nature of Australia: a portrait of the island continent*. New York: Facts on File. Published with the Australian Broadcasting Association.

Attractive, comprehensive view of Australasian nature extending to New Guinea and Hawaii. Includes dibbler, possums, wallabys, and other mammals. See item 99 in **General** Chapter.

385. Veit, P. 1983. Is there new hope for the mountain gorilla? *Defenders* 58(Feb-Apr): 20-27.

Not seen.

386. Volger, G., and S. Volger. 1988. Northern elephant seals. *Sea Frontiers* 34(6)Nov-Dec: 343-347.

Anecdotal account of sighting seals on Pacific Coast at the juncture of

California and Mexico. Useful for its history of the causes of the near extirpation of this species. Amusing description of mating behavior.

387. Ward, G.C. 1987. India's intensifying dilemma: can tigers and people coexist? *Smithsonian* 18(8)Nov: 53-65.

Five out of eight Indian species survive and numbers are starting to increase due to government efforts but villagers are at risk from attack.

388. Weber, M. 1985. Marine mammal protection. *Audubon Wildlife Report* 1985. New York: National Audubon Society. Ed., R. DiSilvestro, 180-211.

389. Weinberg, D. 1986. Decline and fall of the blackfooted ferret. *Natural History* 95(2)Feb: 63-69.

Thought to be extinct until a thriving colony was found in Meeteetse, Wyoming in 1981. Discovery of the ferrets elicited a controversy about taking some of the Wyoming ferrets out-of-state for a captive breeding program.

390. Western, D. 1982. Patterns of depletion in a Kenya rhino population and the conservation implications. *Biological Conservation* 24(6)Oct: 147-156.

391. Western, D., and L. Vigne. 1985. The deteriorating status of African rhinos. *Oryx* 19(4): 215-220.

* White, C.P. 1982. *Endangered and threatened wildlife of the Chesapeake Bay Region: Delaware, Maryland, and Virginia.* Centerville, MD.: Tidewater Publishers.

The Delmarva fox squirrel and cougar are among the mammals mentioned. See item 105 in **General** Chapter.

392. Wiles, G.J. 1987(1988). The status of fruit bats of Guam. *Pacific Science* 41(1-4): 148-157.

Early decline seemingly halted by introduction of fruit bats from Rota. New decline attributed to forest clearing and brown tree snake. On the U.S. Endangered Species List.

393. Williams, D.F., and Kilburn. 1985. Sensitive, threatened, and endangered mammals of riparian and other wetland communities in California. *California riparian systems: ecology, conservation and productive management.* Edited by R.E. Warner and K.M. Hendrix, 950-956. Berkeley, Los Angeles & London: Univ. of California Press.

Species of deer, voles, and mice are considered. Cited in **General** Chapter as item 103.

394. Williams, T. 1986. The final ferret fiasco. *Audubon* 88(3)May: 110-119.

An account of a trip to Meeteetse, Wyoming to see the ferrets. Provides a review of the Fish and Wildlife Service's efforts to preserve this species.

* Wilson, J.H. 1984. *Rare and endangered species of Missouri.* Jefferson City, MO: Missouri Department of Conservation.

Lists 22 mammals. See item 108.

395. Wilson, J.R. 1986. Tourism and gorillas in the Parc National des Volcans, Rwanda. *Primate Eye* No. 19 (Suppl.): 16-19.

396. Wolfheim, J.H. 1983. *Primates of the world: distribution, abundance, and conservation.* Seattle: Univ. of Washington Press.

397. Wood, S.L.. ed. 1986. The black-footed ferret. *Great Basin Nature Memoir* 8. Provo, Utah: Brigham Young University.

398. Woods, C.A. 1981. Last endemic mammals in Hispaniola. *Oryx.* 16(2)Oct: 146-152.

Hutia is suspected to have a larger population in Dominican Republic than counted because its nocturnal habits make sightings rare. However, extinction is threatened in Haiti unless some reserve habitats are set aside. The solenodon is considered "functionally extinct" but author posits it may exist in a remote area.

399. Woodsum, K. 1984. Crisis in wolf country. *Defenders* 59(Jan-Feb): 18-19, 22-27.

Revival of trapping wolves for sport in Minnesota threatens last major wolf population below Canada.

* Woolfenden, G.E. 1983. *Rare, threatened, and endangered vertebrates of southwest Florida and potential OCS (Outer Continental Shelf) activity impacts.* FWS/OBS 82/03. Washington, D.C.: FWS.

Cited in **General** chapter as item 112.

400. Woolley, P.A., and A. Valente. 1982. The dibbler, *Parantechinus apicalis* (marsupialia: dasyuridae): failure to locate populations in four regions in the south of western Australia. *Australian Mammal* 5(4): 241-245.

401. Yom-Tov, Y. 1986. Otters between extinction and survival. *Israel Land and Nature* 11(4): 167-169.

402. ______, and G. Ilani. 1987. The numerical status of *Gazella dorcas* and *Gazella gazella* in the Southern Negev Desert, Israel. *Biological Conservation* 40(4): 245-253.

Survey of 1964-1985 showed *Gazella dorcas* increasing but *Gazella gazella* decreasing. Article discusses possible causes.

BIRDS

* Aiken, S.R., and C.H. Leigh. 1985. On the declining fauna of peninsular Malaysia in the post colonial period. *Ambio* 14(1): 15-23.

Cited in **General** Chapter as item 1.

403. Ali, S., and A.R. Rahmani. 1982-84. Study of ecology of certain endangered species of wildlife and their habitats: The great Indian bustard. *Annual Report, Bombay Natural History Society* 2:1-100.

Publication not found.

404. Amadon, D. 1983. The Everglade Kite. *Florida Field Naturalist* 11(4): 69-72.

Description of a bird considered endangered.

405. Ambuel, B., and S.A. Temple. 1982. Songbird populations in Southern Wisconsin forests: 1954 and 1979. *Journal of Field Ornithology*. 52(2) Spring: 149-158.

Supports studies showing decline in population of long distance migrants when forests are cut down, while short distance migrants survive in forest edge. A discussion of nesting habits.

* Anadu, P.A. 1987. Progress in the conservation of Nigeria's wildlife. *Biological Conservation* 41(4): 237-251.

Cited in **General** Chapter as item 3.

406. Andrew, J.M., and J.A. Mosher. 1981. Maryland's bald eagles. *Maryland Conservation* 57(3): 4-7.

407. Archibald, G. 1984. Cranes: awesome giants of the bird kingdom are being aided by the International Crane Foundation in struggle to survive. *Bulletin of the Field Museum of Natural History* 55(3): 20-3.

Half of the world's fifteen crane species are endangered.

408. Archibald, G., and C.M. Mirande. 1985. Population status and management efforts for endangered cranes. *Transactions of North American Wildlife and Natural Resources* No. 50, p. 586-602.

Analyzes causes of decline in each of 15 species with suggestions for management. Protection of migratory species involves international agreements as breeding areas and winter habitats are widely separated.

409. Armbruster, J.S., ed. 1983. *Impacts of coal surface mining on 25 migratory*

bird species of high federal interest. (FWS Biological Report 83/35). Washington D.C.: U.S. Fish and Wildlife Service.

410. Armitage, D. 1986. World's largest waterfowl staging resounding recovery. *Canadian Geographic* 106(June-July): 46-54.

Hopeful story of the trumpeter swan as a result of its listing as a rare species by the Committee on the Status of Endangered Wildlife in Canada.

411. Backhouse, F. 1986. Bluebird revival. *Canadian Geographic* 106(Apr/May): 33-38.

As part of this account of "the bluebird trail," the setting of specially designed birdhouses by volunteers to replace the disappearing natural nesting sites, the author provides some history of the North American bluebird's distribution and nesting practices.

412. Bain, J.R. 1982. *A profile of the endangered species of Thailand.* Volume 1. *Birds.* Report of the Office of Ecological Services, Florida State Museum No. 4. Gainesville: Office of Ecological Services, Florida State Museum.

413. Baker, W.W. 1983. Decline and extirpation of a population of Red-cockaded Woodpeckers in northwest Florida. In *Red-cockaded Woodpecker Symposium II, Proceedings.* Ed. D.A. Wood, 44-45. Florida Fish and Game Commission, U.S. Fish and Wildlife Service, and U.S. Forest Service, Washington, D.C.

Life history data collected by Tall Timbers Research Station, Florida.

414. Balcomb, K.C.,III. 1982. *Marine birds and mammals of Puget Sound.* Seattle: Sea Grant Program, Univ. of Washington Press.

Cited in **Mammals** Chapter as item 120.

415. Barraclough, P. 1983. Endangered species: Saving California's condors? *Environment* 25(5)June: 2-4.

The San Diego Wild Animal Park's captive breeding program is described along with the issues of this controversial approach. Its critics argue that zoos benefit from the public relations aspect of captive breeding while they would prefer protected habitats and laws passed against shooting and poisoning the condor.

416. Barrows, C. 1986. Cool spots in hot debate. *Living Bird Quarterly* 5(1): 12-16.

The spotted owl is mentioned.

417. Beardsley, T.M. 1988. For the birds: genetic fingerprinting may help to rescue imperiled wildlife. *Scientific American* 258(6)June: 32-34.

Describes a technique for measuring relatedness among species where the population is down to very small numbers. This knowledge is necessary to avoid reproducing recessive genes.

418. Beecher, W.J. 1986. The piping plover: A newcomer to the endangered species list. *Bulletin of the Field Museum of Natural History* 57(3): 24-25.

Describes nesting pattern at beaches around the the Great Lakes. The latest count is but 17 pairs.

419. Beichle, U., and M. Maelzer. 1985. *A conservation programme for western Samoa.* ICBP Technical Publication no. 4. Cambridge, England.

Identifies several endangered species and describes conditions contributing to shrinking numbers of other species in the tropical rain forest areas.

420. Beissinger, S.R. 1986. Demographic environmental uncertainty and the evolution of mate desertion in the snail kite, *Rostrhamus-sociabilis. Ecology* 67(6): 1445-1459.

This endangered hawk has a unique breeding pattern that includes desertion by either mate. Paper is a scientific study of conditions that permit an optimum clutch size for reproduction allowing for mate desertion during the nesting period. Nesting success and population size were found to be related to Florida Everglades water level.

421. _______. 1988. A faithful, fickle hawk. *Natural History* 97(1)Jan: 43-50.

A popular account of the nesting behavior of the snail kite in the Everglades where the current number is 560.

* Bendiner, R. 1981. *The fall of the wild, the rise of the zoo.* New York: E.P. Dutton.

The peregrine falcon is among the endangered described. See also item 10 in **General** Chapter.

422. Berger, J.J. 1988. A big hand for a little owl. *Animal Kingdom* 91(2)Mar/Apr: 28-29, 31.

Endangered in California only, the world's smallest owl is the elf owl. Author's effort to capture and breed Arizona owls in California provides information on mating and nesting.

* Berger, T.J., A.M. Neuner, and S.R. Edwards. 1981. *Directory of federally controlled species.*

Cited in **General** Chapter as item 11.

423. Bhushan, B. 1986. Rediscovery of the Jerdon's or Double-banded Courser, *Cursorius bitorquatus. Journal of the Bombay Natural History Society* 83(1)Aug: 1-14.

First recorded in 1848, sightings ceased early in 1900's. Thought to be extinct, but a new sighting was reported in January 1986.

424. Block, R. 1981. Michigan's endangered species program. *Endangered Species Technical Bulletin* 6(6): 4-6.

News of the Kirtland warbler, bald eagle, peregrine falcon and others. Also cited in **Mammals** Chapter as item 129.

425. Bomholt, R. 1981. *Population trends in Danish raptors since 1970.* Proceedings of the Nordic Congress of Ornithology, no.3, 39-44.

426. Bowles, M.L., K. Kerr, R.H. Thom, and D.E. Birkenholz. 1981. Threatened, endangered, and extirpated birds of Illinois prairies. In *The Prairie Peninsula in the "Shadow" of Transeau: Proceedings of the Sixth North American Prairie Conference, Ohio State University,* August 1978. eds. R.L. Stuckey and K.J. Reese, 257-260. Ohio Biological Survey, Biological Notes No. 15. Columbus, Ohio.

Comments on 16 species that are considered threatened, endangered, or extirpated in Illinois.

* Boyd, J.M. 1984. Nature conservation. *Proceedings of the Royal Society of Edinburgh.*

Puffins, eagles, wrens and others. Cited in **General** Chapter as item 12.

* Branson, B.A., D.F. Harker, Jr., J.M. Baskin, M.E. Medley, D.L. Batch, M.L. Warren, Jr., W.H. Davis, W.H. Houtcooper, and B. Monroe, Jr. 1981. Endangered, threatened, and rare animals and plants of Kentucky. *Transactions of the Kentucky Academy of Science* 42(3-4): 77-89.

Cited in **General** Chapter as item 13.

428. Brittingham, M.C., and S.A. Temple. 1983. Have cow-birds caused forest songbirds to decline? *Bioscience* 35(1)Jan: 31-35.

The brown-headed cowbird is the only obligate brood parasite in North America. They never rear their young and are dependent on host nests. Many songbirds do not have the genetic mechanism to eject foreign eggs or start new nests. This cuts their own reproductive success. Authors studied nests of forest songbirds for evidence of parasitism and distance between nests and opening in forest as cowbird prefers open habitat.

429. Britton, D.G., and T. Hayashida. 1981. *The Japanese Crane: bird of happiness.* Tokyo, New York and San Francisco: Kodansha International.

Describes ICBP's efforts in arranging international migratory bird treaties for preserving the crane, a familiar symbol in Japanese art and culture. Photographs of bird's role in art in addition to behavior are abundant throughout the book.

430. Britton, P.L., S.N. Stuart, and D.A. Turner. 1984. East African endangered species. In *Proceedings of the Fifth Pan African Ornithological Congress, Lilongew, Malawi, 1980.* Ed., J. Ledger. Johannesburg: Southern African Ornithological Society.

Not seen.

431. Brooke, R.K. 1984. *South African red data book-birds*. Pretoria: Foundation for Research Development, Council for Scientific and Industrial Research.

Not seen.

432. Brooke, R.K. 1984. An assessment of rare, vulnerable and endangered South African breeding birds. In *Proceedings of the Fifth Pan African Ornithological Congress, Lilongwe, Malawi, 1980,* ed., J. Ledger. Johannesburg: Southern African Ornithological Society.

Not seen.

433. Brower, K. 1983. The naked vulture and the thinking ape. *Atlantic* 252(4): 70-88.

Preserving the condor has split biologists into two opposing groups: one favors captive breeding and the other proposes protecting the bird in its natural environment. Article reflects intensity of the division, describes positions of scientists working in this field, and reviews the Condor Recovery Program.

434. Brown, J., and D. Guravich. 1983. *The return of the brown pelican.* Baton Rouge: Louisiana State Univ. Press.

Account of the recovery of DDT-affected birds.

435. Brown, P.B., and R.I. Wilson. 1982. The orange-bellied parrot. In *Species at risk—research in Australia.* ed. R.H. Groves and W.D.L. Ride, 106-115. Berlin, Heidelberg, New York: Springer Verlag.

One of four neophoma parrots in Australia, this population is at the critical stage.

436. Brown, P.B., P.R. Wilson, R. Loyn, and N. Murray. 1985. *The orange-bellied parrot.* Royal Australasian Ornithological Union, Report no. 14. Victoria, Australia. (A RAOU Conservation Statement).

Bird has become rare and endangered through intense trapping to market birds as caged pets. Fewer than 200 survive.

437. Brown, R. 1988. Lament for a rare bird: by 2000 AD, the freira will be extinct. *Nature Canada* 17(1): 51-55.

The seabird nests on land making its young easy prey. A small clutch makes its survival unexpected.

438. Bryant, I.J. 1985. Rare birds in the National Wildlife Centre, Mt. Bruce, Masterton, New Zealand. *Avicultural Magazine* 91(1-2): 48-59.

Lists 18 threatened or endangered native birds; each listing accompanied by descriptive paragraph.

439. Buckley, P.A., M.S. Foster, E.S. Morton, R.S. Ridgley, and F.G. Buckley,

eds. 1985. *Neotropical ornithology*. Ornithological Monographs, no. 36. Washington, D.C.: AOU.

Several articles from this volume that describe endangered or threatened status have been individually cited in the present chapter.

440. Bulman, C., D. Rounsevell, and J. Woinarski. 1986. *Forty-spotted pardalote*. Royal Australasian Ornithological Union Report No. 17, Victoria, Australia. (A RAOU Conservation Statement).

Found only in Tasmania, 2500 to 3000 individuals were counted in the last census.

441. Burnham, L. 1987. Threat to the spotted owl. *Scientific American* 257(Oct): 34-36.

Report of a petition to list this bird on the U.S. Endangered List. Presence on the list opens doors to federal funding for preservation in addition to protection of the habitat.

442. Burnside, F.L. 1983. The status and distribution of the red-cockaded woodpecker in Arkansas. *American Birds* 37(2)Mar-Apr: 142-145.

Census reveals a greater number than previously reported on public lands but population is shown to be declining on privately-owned land. Nesting habits show preference for tree cavities not usually found on cleared, privately held land.

443. Burton, P. 1983. *Vanishing eagles*. New York: Dodd, Mead. Illustrated by T. Boyer.

Popular style descriptions of 30 rare species. Illustrations include habitat and food sources.

* Byrne, G. 1988. Strengthened Endangered Species Act passes. *Science* 242(4876)Oct 14: 190.

Cited in **General** Chapter as item 14.

444. Byrne, G. 1988. California, here they go. *Science* 242(4885)Dec. 23: 1639.

A brief update concerning the number of vertebrate classes in California and the awareness of their decline. Describes the California Extinction Index developed by the California Nature Conservancy. See also **Mammal** Chapter, item 149.

445. Cade, T.J., and V.J. Hardaswick. 1985. Summary of peregrine falcon production and re-introduction by the Peregrine Fund in the United States, 1973-1984. *Avicultural Magazine* 91(1-2): 79-92.

Presents useful overview of a well-publicized and successful conservation project begun at Cornell University that involved placing nests and chicks in downtown high-rise buildings.

446. Cade, T.J., J.H. Enderson, C.G. Thelander, and C.M. White, eds. 1988. *Peregrine falcon populations: their management and recovery*. Boise, ID: The Peregrine Fund.

A collection of 81 papers from a symposium held in 1985 in Sacramento, California to commemorate the 1965 conference that connected population decline of the peregrine to organo-chlorine pesticides. The papers cover the status of the peregrine in North America, Europe and elsewhere, migration and banding, and dynamics and ecology. A 'must see' book for anyone concerned with the peregrine.

* Cadieux, C. 1981. *These are the endangered*. Washington, D.C.: Stone Wall Press.

See **General** Chapter, item 15.

447. Cahn, R., and P. Cahn. 1985. Saved but threatened. *Audubon* 87(3)May: 48-51.

Lists IUCN's 12 most threatened animals. Also cited as item 147.

448. Carey, J. 1988. Massacre on Guam. *National Wildlife* 26(5)Aug-Sept: 13-15.

Efforts to save the two remaining endemic birds from extinction brought discovery that the introduction of the brown tree snake caused decline.

449. Carothers, J.H., S.R. Sago, and R.B. Hansen. 1983. Ecological observations on an endangered species: The Maui parrotbill, *Pseudonestor xanthophrys*. *American Birds* 37(5)Sept-Oct: 820-821.

Observations of Hawaii's popular honeycreeper.

450. Carter, J.H., III, R.T. Stamps, and P.D. Doerr. 1983. Status of the red-cockaded woodpecker in the North Carolina pine dunes. In *Red-cockaded woodpecker symposium II, Proceedings*,ed. D.A. Wood, 24-29. Florida Fish and Game Commission, Fish &Wildlife Service, Forest Service, Washington, D.C.

451. Castaldo, N. 1985. A new hope for bluebirds. *Conservationist (New York)* 39(Mar-Apr): 8-11.

A description of the mating and nesting behavior. Bluebirds were a common species early in the 20th century but a 90% decline was noted by the 1960's. The New York Department of Conservation lists the the bluebird as a "species of special concern" and is promoting a network of bluebird bird-houses.

452. Catsadorakis. G. 1986-87. Notes on the avifauna of Lake Prespa National Park. *Biologia GalloHellena* 12:485-496.

Not seen but purportedly describes birds in this area of Greece.

453. Clancey, P.A. 1985. *The rare birds of southern Africa*. Johannesburg: Winchester Press.

Each bird receives a description of its natural history and, in some cases, the circumstances of its identification. Generously illustrated.

454. Clements, J. 1981. *Birds of the world: A checklist*. New York: Facts on File.

Birds that are considered rare are so noted in parentheses in the main body of the work following the specific name of the bird. Organization of the book is for the taxonomically well-informed. Not for the beginner.

455. Coats, S., and W.H. Phelps, Jr. 1985. *The Venezuelan red siskin: case of an endangered species*. Ornithological Monograph no. 36, 977-985. American Ornithological Union. Washington, D.C.

Popular cagebird trade places this finch in danger of extinction.

* Coffin, B., and L. Pfanmuller, eds. 1988. *Minnesota's endangered flora and fauna*. Minneapolis: Univ. of Minnesota Press.

Cranes, falcons, curlews, plovers, owls and many others are among the birds described in this book. See item 17 in **General** Chapter.

456. Cohn, J.P. 1985. Lead shot poisons bald eagles. *Bioscience* 35(90)Sept: 474-476.

Eagles' diet may include lead pellets ingested by their prey. Urges hunters to use steel shot.

457. Cohn, J.P. 1982. The condor besieged: a shared concern. *Americas* 34(6) Nov/Dec: 3-8.

The world's heaviest fully flighted bird has a wingspread over 10 feet. Andean condor being studied to aid researchers in developing conservation strategy for California condor.

458. Coles, D. 1986. Results of the hooded siskin census. *Avicultural Magazine* 92(2): 100-108.

459. Collar, N., and P. Andrew. 1988. *Birds to watch: The ICBP world checklist of threatened birds*. ICBP Technical Publication no. 8. Distributed in the United States and Canada by the Smithsonian Institution Press.

Reports a dramatic increase in the number of species at risk since the last survey of 1978. Contains information about each threatened bird.

460. Collar, N.J., and S.N. Stuart. 1985. *Threatened birds of Africa and related islands*. ICBP/IUCN Red Data Book, Part 1. 3rd ed. Cambridge, UK.

First of 4 regional volumes planned as the 3rd edition of the *ICBP Red Data Book,* prepared by two conservation groups recognized for the authority and meticulousness of their reports. An extensive account of each threatened species and a series of indexes that provide access to the data.

461. Conant, S. 1981. Recent observations of endangered birds in Hawaii's

national parks. *Elepaio* 41(7): 55-61.

462. Conkling, P.W. 1985. They bluffed the puffins. *National Wildlife* 23(5)Aug-Sep: 4-9.

An account of re-establishing the puffin on U.S. territory. Although puffins are among the most numerous seabirds in the northern hemisphere, by 1880 they had all but vanished from Maine coastal islands. Relates efforts of habituating a colony on Eastern Egg Rock, Maine.

463. Connor, J. 1988. Empty skies: Where have all the songbirds gone? *Harrowsmith* 3(16)Jul-Aug: 35-45.

Overview of the global causes of the songbird decline in relation to international migration trends.

464. Connor, R.N., and K.A. O'Halloran. 1987. Cavity tree selection by Red-cockaded woodpeckers as related to growth dynamics of southern pines. *The Wilson Bulletin* 99(3)Sept: 398-412.

Research on nesting habits of endangered woodpecker. Authors state that decline of woodpecker is due to declining habitat and view nesting research as an important step in planning preservation programs.

465. Conover, A. 1988. The macaws of Manu. *Defenders* 63(4)Jul-Aug: 10-16.

The macaws are the largest of the parrot family. Included in the article are good physical descriptions, nesting and mating behavior, identification of predators and causes of endangerment. Article is accompanied by glorious photographs.

466. Conroy, M.J., J.R. Goldsberry, J.E. Hines, and D.B. Stotts. 1988. Evaluation of aerial transect surveys for wintering American black ducks. *Journal of Wildlife Management* 52(4)Oct: 694-703.

Survey made of winter population in New Jersey and along the Atlantic flyway out of concern for the declining population of American black ducks. Paper concerned with developing a methodology for surveys and relates migration patterns as part of the process.

* Convention on International Trade in Endangered Species (CITES). 1981-. *Identification manual.*

Cited in **General** Chapter as item 19.

467. Crawford, M. 1985. The last days of the wild condor? *Science* 229(4716) Aug 30: 844-845.

Another view of the fate of this much-discussed raptor.

468. Crivelli, A.J., H. Jerrentrup, and T. Mitchev. 1988. Electric-power lines – a cause of mortality in Pelecanus-*Crispus bruch*, a world endangered bird species, in Porto-ago, Greece. *Colonial Waterbirds* 7(2): 301-305.

Not seen.

469. Crocoll, S., and J. Parker. 1988. The life and times of the Red-shouldered hawk. *Conservationist (New York).* 43(1)Jul/Aug: 41-45.

Classed as "threatened" in New York, the red shouldered hawk has also been given special status in 17 other states because it is on the decline in its traditional range. The authors are optimistic about the future because of the results of a 3-year study and the establishment of an on-going Breeding Bird Atlas study.

470. Croxall, J.P., P.G. Evans, and R.W. Schreiber, eds. 1984. *Status and conservation of the world's seabirds.* ICBP Technical Publication no. 2, Cambridge, UK.

A collection of papers on seabirds based on the proceedings of the ICBP Seabird Conservation Symposium, Cambridge, England, 1982. Not all the seabirds are officially listed as endangered but many authors argue that more should be.

471. Currey, D. 1987. Senegal's wild bird trade. *Animal Welfare Institute Quarterly* 36(1-2)Spr/Sum: 1-3.

CITES was asked to investigate when the wild bird trade—larger than officially reported—was uncovered in Senegal, the world's largest exporting wild-bird country.

* Cypher, E.A., R.H. Yahner, G.I. Storm, and B.L. Cypher. 1986. Flora and fauna survey in a proposed recreational area of Valley Forge National Historical Park. *Proceedings of the Pennsylvania Academy of Sciences* 60(1): 47-50.

Cited in **General** Chapter as item 23.

472. Darby, J. 1985. The great yellow-eyed penguin count. *Forest Bird* 16(2): 16-18.

Not seen. Cited in ZooRecord.

473. Darlington, D. 1987. *In condor country.* Boston: Houghton Mifflin.

Interviews with long-term residents of the California area that had been home to the condor. The natural history of the area, the condor's relation to humans, and attitudes toward conservation strategies are revealed.

474. Davies, S.J.J.F., G.T. Smith, and F.N. Robinson. 1982. The scrub-bird in Western Australia, In *Species at risk—research in Australia,* eds. R.H. Groves and W.D.L. Ride, 117-127. New York and Berlin: Springer Verlag.

Only one site now observed for this species.

475. Davison, G. 1987. "Interwader" and the conservation of coastal wetlands. *Malayan Naturalist* 40(3/4) Mar: 19-21.

The Interwader Project, begun in 1983, sponsored an intensive study of one group of birds that includes the milky stork considered endangered by the project. The project's purpose is to develop effective and focused conservation plans for a shoreline habitat that is at risk from economic and population pressures.

476. DeIongh, H. 1983. Is there still hope for the Bali mynah? *Tigerpaper* 10(1): 28-32.

477. Delany, M.F., and J.A. Cox. 1986. Florida grasshopper sparrow breeding distribution and abundance in 1984. *Florida Field Naturalist* 14(4): 100-104.

Authors concluded this bird bears watching.

478. DeLotelle, R.S., R.J. Epting, and J.R. Newman. 1987. Habitat use and territory characteristics of Red-cockaded woodpeckers in central Florida. *The Wilson Bulletin* 99(2)June: 202-217.

Research of woodpecker's Florida habitat to ascertain cause of decline of bird and plan its preservation.

479. DeSante, D.F., and P. Pyle. 1986. *Distributional checklist of North American birds.* Vining, CA: Artemisia Press.

Rare and endangered birds are so indicated.

480. Dharmakumarsinhju, K.S. 1982. The great Indian bustard (*Ardeotis nigriceps*) in decline. In *Bustards in decline,* eds. P.D. Boriup and H. Vardhan, 127-132. Jaipur: Tourism and Wildlife Society of India.

From a collection of papers concerned with this endangered family.

481. Diamond, A.W. and T.E. Lovejoy, eds. 1985. *Conservation of tropical forest birds.* ICBP Technical Bulletin no. 4, Cambridge, England. Proceedings of a workshop and symposium held at the 18th World Conference of the ICBP, August 1982, Kings College, Cambridge, UK.

Conference held to bring attention to the plight of tropical forest birds who face a greater threat of extinction than avifauna of other areas largely because of the social and economic pressures on the land use of tropical forests.

482. Diamond, A.W., ed. 1987. *Studies of Mascarene Island birds.* Cambridge (UK) and New York: Cambridge Univ. Press.

Useful reference resource for studies resulting from 1974 British Ornithological Union expedition. Mascarene is the collective name for the islands of Mauritius, Reunion, and Rodrigues, home of unusual species.

483. Diamond, A.W., R.L. Schreiber, D. Attenborough, and I. Prestt. 1987. *Save the birds.* Cambridge, UK: Cambridge Univ. Press.

Written to be persuasive and informative. Primarily about the need for a global perspective of the importance of conserving ecosystems but includes descriptive case histories about threatened species.

484. Diamond, J.M. 1987. Endangered birds that died in our sleep. *Proceedings of the Jean Delacour/IFBP symposium on breeding birds in captivity,* 2nd. San Diego, California. pp. 196-201.

Sponsored by the International Foundation for the Conservation of Birds. This is not to be confused with the ICBP, the International Council for Bird Preservation. See Appendix for address.

485. Diamond, J.M., and C.R. Veitch. 1981. Extinctions and introductions in the New Zealand avifauna. *Science* 211(4481): 499-501.

Explores survival status of native avifauna that is competing with "exotic" introduced species.

* Diamond, J.M. 1984. "Normal" extinctions of isolated populations. In *Extinction,* ed. M.H. Nitecki, 193-246. Chicago: Univ. of Chicago Press.

Cited in **General** Chapter as item 25.

486. Dickson, V. 1983. The Lord Howe Island Group. *Wildlife Australia* 20(2): 8-11.

487. Dinsmore, J.J. 1981. Iowa's avifauna: Changes in the past and prospects for the future. *Proceedings, Iowa Academy of Science* 88(1): 28-37.

* DiSilvestro, R.L., and A.S. Eno, eds. 1986. *Audubon wildlife report 1986.* New York: National Audubon Society.

Cited in **General** chapter as item 26.

* DiSilvestro, R.L., ed. 1987. *Audubon wildlife report 1987.* New York: National Audubon Society. (Distributed by Academic Press.)

Cited in **General** Chapter as item 27.

488. Dobson, A.P. 1988. Restoring island ecosystems: the potential of parasites to control introduced mammals. *Conservation Biology* 2(1)March: 31-39.

The introduction of small mammals, e.g., rats, cats, mongoose, etc., has decimated many an endemic island population, particularly birds. To assist the stabilization of the Laysan albatross in Hawaii, it is proposed that certain parasites be introduced to reduce the population of the intruders.

489. Doughty, R.W. 1983. *Wildlife and man in Texas.* College Station: Texas A & M Univ. Press.

Argues that decline of fish and game during 19th century were due to overhunting. Includes tables of population data for protected species.

490. Duffy, D. 1984. The endangered petrel of the Galapagos volcanoes. *Noticias de Galapagos* 39:24-27.

A difficult article to find.

491. Dumbell, G. 1986. The New Zealand brown teal:1845-1985. *Wildfowl* 37: 71-87.

492. Durrell, G.M. 1983. *Ark on the move*. New York: Coward McCann.

An anecdotal description of rare island animals primarily on Mauritius and other Indian Ocean islands by a well known proponent of animal preservation through zoo breeding. Photographs are from the television series on which the book is based. Also cited in **Mammals** Chapter, item 181.

493. Durrell, G.M. 1984 [1977]. *Golden bats and pink pigeons*. 2nd ed. New York: Simon and Schuster.

Description of the author's efforts to preserve rare animals by maintaining a sanctuary in England on the Isle of Jersey. Cited in **Mammals** Chapter as item 182.

494. Eagles, P.F.J., and J.D. McMauley. 1982. The rare breeding birds of Ontario. *University of Waterloo, Biology Series* no. 24, pp. 1-16.

495. Eddleman, W.R., F.L. Knopf, B. Meanley, F.A. Reid, and R. Zembal. 1988. Conservation of North American rallids. *The Wilson Bulletin* 100(3)Sept: 458-475.

Habitats of rallids are in or near wetlands. Habtat requirement and range is discussed. Determining rarity status in rallids is complicated as rails are often categorized as game birds.

496. Edgecombe, J. 1983. Survival at a price: the case of the Lord Howe Island woodhen. *Habitat Australia* 11(1): 12-13.

* Ehrlich, P.R., and A. Ehrlich. 1981. *Extinction: The causes and consequences of the disappearance of species*. New York: Random House.

A seminal work describing what have become classic cases in extinction. Listed in **General** chapter as item 29.

497. Ehrlich, P.R., D.S. Dobkin, and D. Wheye. 1988. *The birders handbook: a field guide to the natural history of American birds: including all species that regularly breed north of Mexico*. New York: Simon and Schuster.

A compilation of information about the species known to nest in the United States and Canada, roughly 650. Not a guide to identify birds, but a description of each that includes behavior, biology, diet, reproduction, references, miscellaneous notes and, for the purposes of this bibliography, a comment on conservation. Birds on the Blue List and the U.S. Endangered List are cited; birds of regional or local concern are also noted. Interspersed among the species descriptions are wonderfully varied essays on topics such as bird behavior and physiology, biographies of well-known ornithologists, extinct and endangered species. Extensive bibliography: 672-735.

498. Engbring, J., and H.D. Pratt. 1985. Endangered birds in Micronesia: their history, status and future prospects. *Bird Conservation* 2:71-105.

499. Evans, D.L. 1982. *Status reports on twelve raptors*. U.S. Fish and Wildlife Service. Special Scientific Report, Wildlife no. 238, Washington, D.C.

A 68-page description of raptors in the United States that are extinct, endangered or vulnerable.

500. Evans, S., and M. Fidler. 1986. *The Gouldian Finch*. New York, Sydney: Blandford and Poole Press.

Attributes the decline in the wild to popularity as caged birds. Authors explore environmental conditions needed for population maintenance in the wild. Concerned that captivity and inbreeding will create artificial conditions permitting genetic changes.

501. Fefer, S.I., M.B. Naughton, C.S. Harrison, and R.J. Shallenberger. 1984. Synopsis of results of recent seabird research conducted in the northwestern Hawaiian Islands. *Proceedings of the symposium on resource investigations in the Northwestern Hawaiian Islands*. ed. R.W. Grigg and K.Y. Tanoue, 9-76. Honolulu: Univ. of Hawaii Sea Grant Program.

502. Ffrench, R. 1985. *Changes in the avifauna of Trinidad*. Ornithological Monograph no. 36, 987-991. Washington, D.C.: American Ornithological Union.

Describes birds thought to have been extirpated during the 20th century and those no longer observed as breeding pairs.

503. Field, R., and B.K. Williams. 1985. Age of cavity trees and colony stands selected by red-cockaded woodpeckers. *Wildlife Society Bulletin* 13(2): 92-96.

504. Findholt, A. 1986. Pathfinder pelicans. *Wyoming Wildlife* 50(5): 4-7.

505. Fjeldsa, J. 1984. Three endangered South American greves (Podiceps): Case histories and the ethics of saving species by human intervention. *Annales Zoologici Fennici* 21(3):411-416. (In English)

There are natural causes for population declines as well as man-made alterations of the environment. Argues that strategy for manipulating habitats for preservation should be based on the cause of population decline.

506. Forsman, E.D., E.C. Meslow, and H.M. Wright. 1984. Distribution and biology of the spotted owl in Oregon. *Wildlife Monograph* No. 87. Supplement to *Journal of Wildlife Management* 48(2)April. Washington, D.C.: Wildlife Society.

The survival of the spotted owl is of concern in the northwest. Application has been made but it is not yet on the U.S. Endangered List.

* Franklin, K.E. 1987. Endangered species: Where to from here? *American Forests* 93(11-12)Nov/Dec: 57-60, 74-77.

Cited in **General** Chapter as item 33.

507. Franzrib, K.E. 1987. Endangered status and strategies for conservation of the Least Bell's vireo (*Vireo bellii pusillus*) in California. *Western Birds* 18(1): 43-49.

Well organized paper giving natural history, current status of research, analyses of causes of decline and successful conservation efforts.

508. Freeman, J.T. 1984. Woodsman, spare that woodpecker! *Defenders* 59(6)Nov/Dec: 4-13.

509. Freethy, R. 1987. *Auks, an ornithological guide.* New York: Facts on File.

Although auks are not on the U.S. Endangered List, some species have vanished from certain islands where they had once lived. Author devotes a chapter to threats of the modern world on these birds: primarily oil spills, a very current concern.

510. Fullagar, P.J. 1985. The Woodhens of Lord Howe Island. (*Tricholimnas sylvestris*) *Avicultural Magazine* 91(1-2): 15-30.

Perilous state known by 1979, initiating cooperation between research and management groups. Wonderfully descriptive of the terrain and includes a useful table of the status of a variety of birds.

511. Fuller, E. 1988. *Extinct birds.* New York: Facts on File.

What appears to be a coffee table book is truly a "must see" that is more inclusive than the title. Not only a study of birds that have become extinct since 1600 but a review of birds that are rare, endangered, and threatened at present in addition to those with unstable populations. Handsomely illustrated with many color photographs and indexed in great depth. Over 500 species are mentioned.

512. Furness, R.W., and P. Monaghan. 1987. *Seabird Ecology.* New York: Chapman and Hall. London and Glasgow: Blackie and Son.

Primarily about relations between seabirds and man. The last chapter, "Conservation requirements," provides case histories of several bird species whose extinction is complete or nearly so. A good bibliography.

513. Garrido, O.H. 1985. *Cuban endangered birds.* Ornithological Monograph 36:992-999.

Most endangered forms are restricted to keys around Cuba. See item 433 for full reference.

514. Garson, P.J. 1983. The cheer pheasant, *Catreus wallichii*, in Himachal Pradesh, western Himalayas: an update. *Journal of World Pheasant Association* 8: 29-39.

A difficult journal to find. Many pheasants are on the U.S. Endangered Species List.

515. Gasperetti, J., and P. Gasperetti. 1981. Birds of Saudi Arabia: A note on Arabian ornithology—two endangered species. *Fauna Saudi Arabia* 3: 435–440.

Not seen. Difficult journal to find.

516. Gaston, A.J. 1984. Is habitat destruction in India and Pakistan beginning to affect the status of endemic passerine birds? *Journal of the Bombay Natural History Society* 81(3)Dec: 636-641.

Suggests passerine survival more successful in Ganges than in Indus basin. Surveys reveal relation between intensity of ecosystem alteration and proportion of survival of local species.

517. Gerrard, J.M., and G.R. Bortolotti. 1988. *The bald eagle: Haunts and habitats of a wilderness monarch*. Washington, D.C. and London: Smithsonian Institution Press.

Observation of the breeding pattern and development of a bald eagle family in a sanctuary/research center on Besnard Lake, Saskatchewan. Extensive Bibliography: 166-175.

518. Giffen, J.G., J.M. Scott, and S. Mountainspring. 1987. Habitat selection and management of the Hawaiian crow. *Journal of Wildlife Management* 51: 485-494.

519. Gill, F. 1985. Birds. In *Species of special concern in Pennsylvania*, eds. F. Genoways and F. Brenner. Carnegie Museum of Natural History, Special Publication No. 11. Pittsburgh.

Compiled as part of the Pennsylvania Biological Survey, lists endangered and threatened species by taxonomic order. See also **General** Chapter, item 35.

520. Gnam, R. 1987. Underground parrots. *Animal Kingdom* 90(5)Sept/Oct: 41, 43-44.

Once an inhabitant of seven Bahamian islands, this parrot is now found on only two. Its underground nesting pattern is unusual among new world parrots. Article describes breeding and care of the young.

521. Gollop, J.B., T.H. Barry, and E.H. Iversen. 1986. *Eskimo curlew: a vanishing species?* Saskatchewan Natural History Society. Special Publication No. 17.

Complete history of this rare bird that has always seemed to be on the verge of extinction. Last sighted in 1981 in Texas, but nests have not been seen since 1866. Excellent bibliography lists 600 references.

522. Goriup, P.H., ed. 1988. *Ecology and conservation of grassland birds*. ICBP Technical Publication no. 7.

523. Goriup, P.H., and H. Vardharn, eds. 1983. *Bustards in decline*. Proceedings of the International Symposium on Bustards, Nov. 1-3, 1980. Jaipur: Tourism and Wildlife Society of India.

Contains over 50 papers concerned with the case histories, ecology, and conservation of this rare bird family of which there are but 22 species left.

524. Goudie, A. 1986. *The human impact on the natural environment*. 2d ed. Cambridge: MIT Press.

Case histories of several extinctions are part of the discussion of policies that cause environmental decline. Predicts increasing rate of species extinction as the result of diminished areas of natural habitat. More important for its theoretical approach than for information about individual species. Flightless rail and dodo are mentioned.

525. Graham, F., Jr. 1988. A bird in the hand is worth ...plenty! *Audubon* 90(2)March: 96-99.

Indonesia and the New York Bronx Zoo are in a joint program that breeds rare Indonesian species in the Bronx zoo and returns birds to Indonesia. As birds become rare their price in the pet market increases.

526. Graham, F., Jr. 1983. A silence on Guam. *Audubon* 85(5)September: 12-15.

The guam rail (or Koho) is the only native surviving rail on the island. Author criticizes U.S. Fish and Wildlife Service for ignoring Micronesia.

527. Graham, F., Jr. 1986. Cry of the plover. *Audubon* 88(3)May: 12,14,16-17.

A shore bird in conflict with water resources development.

528. Gray, M.V., and J.M. Greaves. 1984. Riparian forest has habitat for the least bell vireo. In *California riparian systems: ecology, conservation and productive management,* eds. R.E. Warner and and K.M. Hendrix, 605-611. Berkeley, Los Angeles, and London: Univ. of California Press.

Vegetation conditions relevant to breeding success. Cited in **General** Chapter as item 103.

529. Gray, P. 1983. The New Zealand plover—the forgotten one. *Forest Bird* 14(5): 27-28.

Not seen.

530. Gray, P. 1981. The black stilt is slipping away. *Forest Bird* 13(219): 26-27.

531. Greed, G. 1984. The white stork (*Ciconia ciconia*) a review of the species with special reference to breeding in the United Kingdom, especially at the Bristol Zoo. *Proceedings of the Symposium of the Association of British Wild Animal Keepers,* no. 9, pp. 4-10.

532. Griffen, C.R. 1985. Biology of the Hawaiian hawk, *Buteo solitarius* (reproduction, food habits, behavior, movements, taxonomy.) *Dissertation Abstracts International* 46(11-B): 3660-3898.

Comprehensive examination of a hawk now estimated to number from 1,400 to 2,500.

* Groves, R.H., and W.D.L. Ride, eds. 1982. *Species at risk: Research in Australia.* Proceedings of a Symposium on the biology of rare and endangered species in Canberra, Australia, November 25 and 26, 1981. Canberra: Australian Academy of Science. New York: Springer-Verlag.

Golden shouldered parrot, pardalote, paradise parrot, wren and bristlebird are among the species noted. See **General** Chapter, item 37.

533. Gutierrez, R.J., and A.B. Carey, Tech. Eds. 1985. *Ecology and management of the spotted owl in the Pacific Northwest.* U.S. Forest Service, General Technical Report, 185. Portland: U.S. Department of Agriculture, Pacific Northwest Forest and Range Experiment Station.

Due to concern for the spotted owl in the northwest, a symposium was held during the 54th annual meeting of the Cooper Ornithological Society at Arcata, California in June 1984. Nineteen papers organized around the issues of management, research and theory of the decline of the spotted owl are contained in this governmental publication.

534. Hambler, C., K. Hambler, and J.M. Newing. 1985. Some observations on *Newillas aldabranus*, the endangered brush warbler of Aldabra atoll, with hypotheses on its distribution. *Atoll Research Bulletin* 290: 1-19.

First described in 1967, the brush warbler is considered the world's rarest and most threatened bird.

535. Hambler, C., K. Hambler, and J. Newing. 1983. The Cambridge Aldabra rail and brush warbler expedition. *Cambridge Expedition Journal* (England) 1983: 72-77.

Expedition to reassess the security of these two endemic, endangered birds found no significant retraction of range in these western Indian Ocean islands since 1973.

536. Hamel, P.B. 1986. *Bachman's Warbler: A species in peril.* Washington, D.C.: Smithsonian Institution Press.

Although single sightings have been reported up to 1980, the bird may be extinct as the last photograph was taken in 1958. Debate on its existence continues in this comprehensive treatise that includes an exhaustive literature search.

537. Hamel, P.B., and S.A. Gauthreaux, Jr. 1982. Field identification of Bachman's warbler (*Vermivora bachmanii* Audubon.) *American Birds* 36(3)May: 235-240.

Provides detailed description of bird's varied plumage. Argues that non-sighting is due to lack of awareness of variety of bird's plumage.

538. Harding, D. 1987. From beyond the horizon (snowy owls). *Conservationist* 41(Mar-Apr): 20-21.

Not seen.

539. Hardy, J. 1984. Endangered birds in New South Wales. *Bird Observer* 627: 34-35.

Problems in Australia, a continent interesting because of its isolation.

540. Harlow, R.F., R.G. Hooper and M.R. Lennarts. 1983. Estimating numbers of red-cockaded woodpecker colonies. *Wildlife Society Bulletin* 11(4): 360-363.

Authors developed what they call "a simple and efficient technique for estimating breeding populations of red-cockaded woodpecker."

541. Harris, J.H., S.D. Sanders, and M.A. Flett. 1987. Willow flycatcher surveys in the Sierra Nevada. *Western Birds* 18(1): 27-36.

Up for consideration by California as endangered or threatened, author suggests brown-headed cowbird is causing decline.

542. Harris, J.T. 1981. *The peregrine falcon in Greenland: observing an endangered species*. Columbia: Univ. of Missouri Press.

Classic style of nature writing: observation of the bird and the people he met during a summer in Greenland.

543. Harrison, H.H. 1984. *Wood warblers' world.* New York: Simon and Schuster.

Anecdotal description of wood warblers only two of which are considered endangered: Bachman's and Kirtland's warblers. Both have received more extensive treatment in other works cited here.

544. Harwood, M. 1983. A clannish problem. *Audubon* 85(4): 30-32.

Examines natural history of red-cockaded woodpeckers and conflict with timber industry.

545. ______. 1986. You can't protect what isn't there. *Audubon* 88(6)Nov: 109-123.

Argues against "myths" setting criteria for endangered status. Notes sighting of ivory-billed woodpecker in Cuba.

546. Hawegawa, H., and A.R. Degange. 1982. The shorttailed albatross, *Diomedea albatrus,* its status, distribution, and natural history. *American Birds*

36(5)Sept: 806-814.

547. Hay, R. 1986. Birds at risk in the Pacific. *Forest Bird* 17(2): 7-9.

548. ______. 1984. The kokako: perspective and prospects. *Forest Bird* 15(1): 6-11.

Not seen. Not to be confused with the kakapo.

549. Hay, R., and G. McCormack. 1984. Conserving the kakirori. *Forest Bird* 15(3): 23-24.

550. Haynes, A.M. 1987. Human exploitation of seabirds in Jamaica. *Biological Conservation* 41(2): 99-124.

Lists of endangered and threatened birds.

551. Heinrichs, J. 1984. Rare birds and big trees: west. The owl and the old growth. *American Forests* 90(3)Mar: 22-23, 54-56.

Northern spotted owl likes old growth forests; forests untouched for 150 years. Spotted owl's habitat in conflict with lumber interests.

552. ______, and D.B. Heinrichs. 1984. Rare birds and big trees: east. The woodpecker and the pines. *American Forests*. 90(3)Mar: 24-26, 46, 48-49.

A very clear explanation of the unusual roosting and breeding patterns of the red-cockaded woodpecker. This is the only woodpecker that prefers live old trees. Live trees ooze resin as birds peck out cavity for nest. Resin traps tree snakes before they can reach the nest. Authors present clear explanation of the clan and colony nesting behavior.

553. Henderson, A.K. 1985. From zero to seventy-five: a vigil at a least tern nesting ground. *Sierra*. 70(Jul-Aug): 79-81.

An endangered colony of California least terns in the San Diego area is preserved by volunteers from the Cetacean Society and an educated, willing public.

554. Henderson, C.L. 1988. Nongame bird conservation. In *Current Ornithology No. 5,* ed., R.F. Johnston. New York and London: Plenum Press.

Argues that wildlife conservation efforts focused on game before 1975, ignoring a broad spectrum of bird life that did not get listed as endangered. Listing promotes government funding for restoration projects, e.g., bald eagle and peregrine falcon. Suggests that periodic surveys of populations will set off alarms about declining numbers. Programs that address the issue during the "decline" stage are more economically efficient than recovery programs.

555. Henke, R. 1988. Mynahs, macaws and money; New York's Wild Bird Act controls the importation of rare and endangered birds. *Conservationist* (New York) 42(Jan-Feb): 26-32.

The title of this article describes its contents fully. Mynahs and macaws and parrots have become popular as caged birds. Although they are on the CITES and the U.S. Endangered Species List, smugglers continue to try to supply buyers.

556. Herman, S.G., J.B. Bulger, and J.B. Buchanan. 1988. The snowy plover in southeastern Oregon and western Nevada. *Journal of Field Ornithology* 59(1): 13-16.

Put on the Blue List in 1971, the authors found the population healthy in a 1980 survey. Only serious threats are diversion of water for irrigation, high water conditions, and lowered water tables due to geothermal activity.

557. Hilty, S.L. 1985. Distributional changes in the Colombian avifauna: a preliminary blue list. *Ornithological Monograph* 36:1000-1012.

The "blue list" is the AOU "endangered" list. For full citation see Buckley et al., item 439.

558. Hindman, L.J. 1985 Trumpeter swan blasts back. *American Birds* 39(3)Fall: 260-264.

Since 1971 the birds are a protected species.

559. Hockey, P.A.R. 1986. Meade-Walso's black oyster catcher—a path to extinction. *African Wildlife* 40(4): 140-151.

560. Hooper, R.G., and M.R. Lennartz. 1983. Roosting behavior of Red-cockaded woodpecker clans with insufficient cavities. *Journal of Field Ornithology* 54(1): 72-76.

Behavior of a recognized endangered bird showing importance of clan during roosting.

561. Hope-Jones, P., and A.W. Colling. 1984. Breeding and protection of Montagu's harriers in Anglesey, 1955-64. *British Birds* 77(2): 41-46.

562. Houston, C.S., and M.J. Bechard. 1984. Decline of the ferruginous hawk in Saskatchewan. *American Birds*. 38(2)Mar-Apr: 166-170.

North America's largest buteo has disappeared from its former range although previous nesting sites were observed. Good bibliography.

* Howard, A., A. Perry, P. Herring, and J. Scott. 1984. *Endangered and threatened species on U.S. Air Force installations*. U.S. Fish and Wildlife Service/Office of Biological Services (FWS/OBS 84/110.) Washington, D.C.

Publication consists of tables listing the vertebrates on U.S. Air Force bases that are classified as endangered, threatened, rare, protected and of special concern both on Federal and State lists. Cited in **General** Chapter as item 42.

563. Huber, K.R. 1982. *Kirtland's warbler: an annotated bibliography, 1852-1980.* Ann Arbor: University of Michigan, Museum of Zoology.

564. Hussain, S.A. 1985. Status of blacknecked crane in Ladakh, 1983: problems and prospects. *Journal of the Bombay Natural History Society* 82(3)Dec: 449-458.

Rejects proposal to collect crane eggs from India for breeding at Wisconsin's International Crane Foundation.

565. Hutchins, B.R. 1985. The status in captivity of four Australian parrots. *Avicultural Magazine* 91(1-2): 63-75.

Considered endangered in the wild, author describes their abundance and location in captivity and includes information about their earlier status.

566. Ilani, G., and B. Shalmon. 1984. Still Negev lappet-faced vultures. *Israel land and nature.* 9(4):168.

A bird expected to have become extinct by now, continues to survive, but barely.

* International Convention Advisory Commission. 1981. *Convention on International Trade in Endangered Species of Wild Fauna and Flora.* CITES.

See **General** Chapter, item 43.

567. International Council for Bird Preservation. 1986. *Bird conservation priorities in Nigeria.* ICBP Study Report no. 12.

568. International Federation for the Conservation of Birds. 1987. *Proceedings of the Jean Delacour/ IFCB symposium on breeding birds in captivity, 2nd, San Diego, California.*

Please note that the IFCB is to be differentiated from the ICBP. See directory for addresses. See also item 484.

569. Ishihara, T. 1982. Some noteworthy insects and birds of Japan in the natural environment growing worse. *Memoirs Coll. Agric. Ehime Univ.* 27(3): 153-170.

570. Istvan, S. 1982. Present situation in great bustard (Otis. T. Tarda) in Hungary. In Goriup and Vardhan (eds.) *Bustards in decline.* For full citation see no. 523.

571. Jackson, J.A. 1986. Why do Red-cockaded woodpeckers need old trees? *Wildlife Society Bulletin* 14(4): 318-322.

A scientific methodology is used in this article although topic is one also discussed in popular journals.

572. Jenkins, J.M., and C.F. Aguon. 1981. Status of candidate endangered bird species on Saipan, Tinian and Rota of the Mariana Islands. *Micronesia* 17(1-

2): 184-186.

573. Johns, A.D. 1987. The use of primary and selectively logged rainforest by Malaysian hornbills (Bucerotidae) and implications for their conservation. *Biological Conservation* 40(3): 179-190.

Authors find that selectively logged forests can support a hornbill population with the fruit they need.

574. Johnsgard, P.A. 1984. *Cranes of the world.* Bloomington: Indiana Univ. Press.

Natural history of 14 cranes (5 of which are endangered or vulnerable) noting subspecies as well. A chapter on endangered species and conservation is quite helpful.

575. Johnsgard, P.A. 1988. *North American owls: biology and natural history.* Washington, D.C.: Smithsonian Institution Press.

The U.S. Endangered List does not include any American owls but there is a movement to enter the North American spotted owl on the list. Neither the U.S. nor the ICBP lists include any North American owls. The final section of each species description is subheaded 'Evolutionary relationships and conservation status' in which population abundance is noted.

575a. ______. 1988. *The quails, partridges, and francolins of the world.* New York: Oxford Univ. Press.

One quail, Merriam's Montezuma, is on the U.S. Endangered List. Book is a fine comprehensive work.

576. ______. 1986. *The pheasants of the world.* New York: Oxford Univ. Press.

Fourteen pheasant species are cited on the U.S. Endangered List and are among the 49 existing species covered in this useful volume.

577. Johnson, W.W. 1983. California condor: Embroiled in a flap not of its own making. *Smithsonian* 14(9)Dec: 73-82.

Explains the two major management controversies: collecting eggs for zoo breeding until chicks can be placed in the wild or creating natural habitat sanctuaries and keeping human hands off.

578. Johnstone, G.W. 1985. Threats to birds on subantarctic islands, 101-121. *ICBP Technical Publication No. 4.*

579. Jolly, A. 1982. Island of the dodo is down to its last few native species. *Smithsonian* 13(June): 94-103.

Status of species on Mauritius.

580. Jolly, J. 1985. Little spotted kiwi: paradise regained for paradise lost? *Forest Bird* 16(1): 15-17.

581. Jones, C.G., and A.W. Owadally. 1981. The world's rarest falcon: the Mauritius kestrel. *Falconer* 7(5): 322-327.

582. Jones, G.R.D. 1981. The cape vulture—an endangered species. *Redwing* pp. 42-46.

583. Judge, R.P., R. Strait, and W.F. Hyde. 1984. Economics of endangered species management: the red-cockaded woodpecker. *Transactions, 49th North American Wildlife and Natural Resources Conference.*

Study of costs connected with meeting the targeted population goal of one colony per 200 acres in North Carolina.

584. Kalla, P.I., and F.J. Alsop, III. 1983. The distribution, habitat preference, and status of the Mississippi kite in Tennessee. *American Birds* 37(2): 146-149.

585. Kantor, M. 1986. The plot to banish woody woodpecker. *Defenders* 61(1)Jan-Feb: 12-19.

Describes the role of snags in old growth forests and the animals that use them as the source of nests and safety from predators.

586. Karl, B.J., and H.A. Best. 1982. Feral cats on Stewart Island; their foods, and their effects on Kakapo. *New Zealand Journal of Zoology* 9(2): 287-294.

The kakapo, also known as the owl-parrot, is on the U.S. Endangered List.

587. Kasparek, M.M. 1986. On the occurrence of the bittern (*Botaurus stellaris*) in Turkey. *Zoology Middle East* 1:33-41.

588. Kear, J. 1985. The Hawaiian goose (*Branta sandvicensis*) in the 1980's. *Avicultural Magazine* 91(1-2): 102-103.

Popularly called the nene (particularly in crossword puzzles), this bird is doing well in captivity but not in the wild.

589. Keith, J.O. 1983. Brown pelicans—can they survive? *Oceanus* 26(1): 62-67.

Not seen.

590. Kelsey, M., and T. Langton. 1984. The endangered birds of Sokoke Forest. *Swara* 7(5): 27-29.

Lists declining bird species in this African forest.

591. Kennedy, R.S. 1983. Can the Filipinos learn to love this bird? *International Wildlife* 13(July): 26-34.

Conflict about preserving the Philippine eagle.

592. Kerlinger, P., and C. Doremus. 1981. Habitat disturbance and the decline of dominant avian species in pine barrens of the northeastern United States.

American Birds 35(1)Jan: 16-20.

Pine and prairie warblers extirpated in Albany preserve. The decline of the warblers, brown thrasher and ovenbird in other preserves are analyzed. Posits that changes in habitat affect success of competing species.

* Kiew, B.H. 1985. Malaysia's 10 most endangered animals, plants, and areas. *Malayan Naturalist.* 38(4)May: 2-3.

A brief paper that lists the author's selection of the most, but not the only, endangered Malaysian animals. It was prepared in response to the IUCN list of the 12 most threatened animals which this author felt was inadequate for Malaysia.

Cited in **General** as item 51.

594. King, C. 1984. *Immigrant killers: introduced predators and the conservation of birds in New Zealand.* Auckland and Melbourne: Oxford Univ. Press.

Argues that validity of protection measures should be based on the causes of population decline, i.e., due to natural evolution or the result of human activity. Traces New Zealand natural history from paleontologic era to Polynesian period to European settlement and introduction of fauna.

595. King, J.F. 1987. Nature reserves of the USSR. *Sierra* 72(3)May/June: 38-45.

Cranes, geese, parrotbills and owl-fish are among the birds described. Animals protected by Soviet conservation policy noted. Article based on 1985 update of 1974 inventory of rare and endangered species published as *The Red Book (Krasnaya kniga).* It follows IUCN classification system. Also cited as item 245.

596. King, J.G., and B. Conant. 1981. The 1980 census of trumpeter swans on Alaskan nesting habitats. *American Birds* 35(5)Sept: 789-793.

Trumpeter swan off the endangered list and doing well in Alaska.

597. King, W.B., comp. 1981. *Endangered birds of the world.* 3rd rev. *The ICBP Bird Red Data Book.* Reprinted text of Volume II, *Aves* in the IUCN Red Data Book Series. Originally published in looseleaf format in two parts, 1978 and 1979 in Morges, Switzerland. Washington, D.C.: Smithsonian Institution Press and the ICBP.

Lists the world's rare and endangered species. Lists birds thought to have become extinct since 1600. The IUCN and the ICBP are non-governmental organizations whose publications have world-wide credibility.

598. King, W.B. 1981. World's rarest birds. *International Wildlife* 11(Sept-Oct): 12-19.

599. Konrad, P.M. 1982. Whooping cranes in North Dakota. *North Dakota Outdoors* 45(3): 2-4.

Not seen.

600. Krol, W. 1983. Status of eagles in Poland. *Bulletin of the World Working Group Birds of Prey* 1:61-66.

Not seen.

601. Kushlan, J.A., and P.C. Frohring. 1986. The history of the southern Florida wood stork population. *Wilson Bulletin* 98(3): 368-386.

A review of nesting sights and habitats. Concern for population stability.

602. Labastille, A. 1983. Drastic decline in Guatemala's giant pied-billed grebe. *Environmental Conservation* 10(4): 346-348.

603. ______. 1983. Goodbye, giant grebe? *Natural History* 92(2)Feb: 64-72.

This and item 602 concern the same bird.

604. ______. 1985. Resplendent quetzel. *Living Bird Quarterly* 4(1)Winter: 27-30.

Text about the habitat of the quetzel (sometimes called quetzal) accompanied by splendid photos.

605. Labisky, R.F., and M.L. Porter. 1984. *Home range and foraging habitat of Red-cockaded woodpeckers in north Florida*. Unpublished report. Gainesville: University of Florida, School of Forest Resources and Conservation, Dept. of Wildlife Range Science.

606. Lambert, F. 1985. The St. Vincent Parrot; an endangered Caribbean bird. *Oryx* 19(1): 34-38.

607. Lande, R. 1988. Genetics and demography in biological conservation. *Science* 241(4872)Sept 16: 14551460.

While some information about two species is included (the spotted northern owl and the red-cockaded woodpecker), the article is important as a theoretical examination of the approaches to the determination of viable populations; comparing the importance of population genetics and demography. Argues that demographic factors may be of more immediate importance in planning for species survival of wild populations but a balance of both are "applicable to species in natural environments." Fine literature review and bibliography. Also cited as item 52.

608. Lavee, D. 1988. Why is the *Houbara chlamydotis undulata* macqueenii still an endangered species in Israel. *Biological Conservation* 45(1): 47-54.

In spite of a conservation program, author suspects disturbed nests are shaking up what had become a stable population.

609. Laycock, G. 1982. Roger's "last" bird. *Audubon* 84 (3)May: 26,28-31.

Anecdotal account of Roger Tory Peterson's finding the lesser prairie chicken, the last bird needed for his life list of native breeding birds, safe only in Texas now.

610. Laycock, G.D., and C. Hanel. 1987. Survival of African penguins (*Sphenis-cus demersus*) at Dyer Island, southern South Africa. *Journal of Field Ornithology* 58(3)Summer: 284-287.

Breeding birds increasing on Dyer Island but survival rate is low in other areas.

611. Laymon, S.A. 1987. Can the western subspecies of the yellow-billed cuckoo be saved from extinction? *Western Birds* 18(1): 19-25.

Destruction of riparian forests and awareness of the noticeable decline of the cuckoo began a concerted effort to document population and nests to determine reintroduction into suitable habitat and the viable breeding population size. Also noted are thinning shells which may be caused by pesticides used in South America although banned in the U.S.

612. Lennartz, M.R., P.H. Geissler, R.F. Harlow, R.C. Long, K.M. Chitwood, and J.A. Jackson. 1983. Status of Red-cockaded woodpecker populations on federal lands in the south. In *Red-cockaded woodpeckers, Symposium II, Proceedings*. ed. D.A. Wood, 7-12.

See item no. 799 for full citation.

613. Leshem, Y. 1984. The rapid population decline of Israel's lappet-faced vulture, *Torgus tracheliotus negevensis*. *International Zoo Yearbook* 23: 41–46.

614. ______. 1985. Vultures under high tension. *Israel Land and Nature* 10(4): 149–153.

615. Lewis, R.E. 1984. The plight of the Philippine eagle. *Annual Report of the Hawk Trust* 14:18-21.

Issued by a British organization, the report was not available in United States libraries that enter holdings into the Online Computer Library Center, OCLC.

616. Ligon, J.D., P.B. Stacey, R.N. Conner, C.E. Bock, and C.S. Adkisson. 1986. Report of the American Ornithologists' Union Committee for the Conservation of the red-cockaded woodpecker. *Auk* 103(4): 848-855.

Dependent on mature open pine woodlands, the red-cockaded woodpecker is classified as endangered because of deforestation in southeastern U.S. The bird's very specialized habitat makes conservation a challenge.

617. Lokki, J., and K. Eklof. 1984. Breeding success of the Red-throated diver (*Gavia stellata*) in southern Finland. (English). In *Proceedings of the Fourth Nordic Ornithological Congress, 1983*, ed. T. Stjernsberg, and Y. Haila. Published in *Annales Zoologici Fennici* 21(3): 417-419.

The red-throated diver is considered an endangered species in Finland and of concern in Sweden. Authors devised system of "ringing" to study breeding patterns. Islet nests appear more successful than nests on shores.

618. Loucks, B.A., and K. Kogut. 1985. The peregrine returns. *Conservationist (New York)* 39(5)Mar/Apr: 12-17.

Provides background of the DDT-induced decline of the hawk and the dramatic efforts for its recovery including some accounts of daring attempts to put new chicks in difficult to reach nests, e.g., girders of the Verrazano Bridge.

619. Low, R. 1984. *Endangered parrots*. Poole, England: Blandford.

Not seen.

620. ______. 1985. Breeding the Tahiti blue lory (*Vini peruviana*). *Avicultural Magazine* 91(1-2): 1-14.

Found only on the Society Islands other than Tahiti, the largest of the group, where it is now extinct. The controversy over its status on the other islands is given.

621. Luoma, J.R. 1987. Black duck decline: an acid rain link. *Audubon* 89(3)May: 19-24.

Postulates that acid rain in lakes is causing decline.

622. Ma, Y., L. Jin, A. Jin, and C. Fu. 1987. An aerial survey on red-crowned cranes and other rare waders in the Wuyuerhe River Basin of Heilongjian Province. *Acta Zoologica Sinica* 33(2)June: 187-191.

Abstract in English. Text in Chinese. Summarizes aerial method of conducting census. Cites numbers found within species.

623. Mackey, D.J. 1982. To help hooked sea birds. *Oceans* 5(3)May-June: 29-35.

News of a sanctuary for mending injured pelicans. The most common cause of unnatural death and injury is from fish hooks and monofilament fish lines. Florida fisherman have killed pelicans and destroyed eggs and chicks in the rookeries because of fear of competition.

* Majumdar, S.K., F.J. Brenner, and A.F. Rhodes, eds. 1986. *Endangered and threatened species programs in Pennsylvania and other states: causes, issues, and management*. Harrisburg: Pennsylvania Academy of Science.

Chapter on birds is relevant. Cited also in **General** Chapter as item 54.

624. Mallinson, J.J.C., and K.M. Taynton. 1983. A checklist of white-eared pheasants (*Crossoptilon crossoptilon*) held in captivity during 1981/1982. *Dodo* No. 20:66-72.

Efforts to breed the pheasant (a Chinese native) in captivity.

625. Manry, D.E. 1986. Ibises in peril. *Living Bird Quarterly* 5(2): 24-27.

626. Marinkovic, W. et al. 1985. The griffon vulture in Yugoslavia. *ICBP Technical Publication* no. 5, 131-135.

627. Marshall, D.B. 1988. *Status of the marbled murrelet in North America: with special emphasis on populations in California, Oregon, and Washington.* U.S. Fish and Wildlife Service, Biological Report 88(30). Washington, D.C.: Fish & Wildlife Service

628. Martin, L. 1983. Crackers over polly. *Americas* 35(Sept-Oct): 40-46.

 Parrot popularity as a caged bird increasing despite illegality of trade under CITES. Includes natural history of bird.

629. Martindale, J. 1985. *The freckled duck.* Royal Australasian Ornithological Report No. 22. (A RAOU Conservation Statement).

 Listed among the world's 10 rarest waterfowl.

630. Massey, B.W. 1981. A least tern makes a right tern. *Natural History* 90(11)November: 62-71.

 Imperiled California bird may have been saved by the Endangered Species Act. Report describes the breeding behavior of the *Sterna antillarum browni.*

* Matthiessen, P. 1987. *Wildlife in America* New York: Viking Penguin. Revised and updated edition. Previously published in 1958.

 See **General** chapter, item 56.

631. May. J. 1981. The last flight of endangered birds. *Wildlife* 23(9): 29-30.

632. Mayfield, H.F. 1984. Kirtland's warbler: why so rare? *Living Bird Quarterly* 3(2)Spring: 4-8.

 Author describes very circumscribed ecological niche the species demands. Interesting ecological balance between nest sites, small jack pines, and forest fires.

633. McCafferty, J. 1987. The ivory-billed woodpecker. *American Forests* 93(11-12)Nov-Dec: 59-61.

 An account of the search for the woodpecker by Jerome Jackson who believes they may yet be found in the U.S.

634. Melville, D.S. 1984. Seabirds of China and the surrounding seas. *ICBP Technical Publication No. 2,* 501-511.

635. Meredith, C.W. 1984. *The ground parrot.* Royal Australasian Report No. 10. (A RAOU Conservation Statement).

 First of a series of monographic reports highlighting a single endangered species.

636. Meyers, J.P. et al. 1987. Scarce resources and common species. *Endangered Species Technical Bulletin Reprint* 4(6): 1-4. University of Michigan.

 Describes the conditions of migratory seabirds.

637. Middleton, H. 1985. The blue robin hangs on. *South Living* 20(June): 54-56.

Not seen.

638. Mills, J., and G. Williams. 1984. New Zealand, New Guinea and Lord Howe Islands: the status of endangered New Zealand Birds. In *Vertebrate zoogeography and evolution in Australasia (Animals in Space and Time).* ed. M. Archer and G. Clayton, 1107-1120. Carlisle, Australia: Hesperian Press.

639. Millsap, B.A. 1986. Status of wintering bald eagles in the conterminous 48 states. *Wildlife Society Bulletin* 14(4): 433-440.

640. Mindell, D.P. 1986. *Nesting raptors in southwestern Alaska: status, distribution, and aspects of biology.* Anchorage: U.S. Department of Land Management.

* Mittermeier, R.A. 1981. Brunei protects its wildlife. *Oryx* 16(1)May: 67-70.

Hornbills and the great argus pheasant. See item 57 in **General** Chapter.

641. Mohammad Ali, S. 1981. The great Indian bustard. *Tigerpaper* 8(4): 9-11.

642. Montevecchi, W.A. and L.M. Tuck. 1987. *Newfoundland birds: exploitation, study, conservation.* Publication of the Nuttal Ornithological Club, No. 21, ed. R.A. Paynter, Jr. Cambridge, Massachusetts.

Argues that many endangered seabirds are no longer breeding in other North Atlantic areas but are nesting in Newfoundland. Optimistic about survival there as long as protection continues. Appendices contain scientific, common and vernacular names, and national parks and sanctuaries.

643. Montgomery, S. 1986. Will the kakapo dance again? *International Wildlife* 16(Nov-Dec): 12-15.

Scientists try to save a parrot noted for its unusual mating calls.

644. Moors, P.J., ed. 1988. Conservation of island birds. *ICBP Technical Publication No. 3*, Princeton: Princeton Univ. Press.

Includes lists of many birds.

645. Morrison, A. 1982. A dissenting view on aviculture. *Avicultural Magazine* 88(2):113-116.

The author views the horned parrot as critically endangered even if it is not on any "list" and argues against its captivity at all.

646. Morrison, M.L. 1981. Population trends of the Loggerhead shrike in the United States. *American Birds* 35(4)July: 754-757.

Confirms species "blue listing" in the east and southeastern United States coasts, but stabile west of the Mississippi.

647. Moulton, D.W., and M.W. Weller. 1984. Biology and conservation of the Laysan duck. *Condor* 86: 105-117.

An endangered Hawaiian bird.

648. Mountfort, G. 1986. Following the dodo? Vanishing birds of the world. *Country Life* 180(4640): 298-299.

Not seen.

* Mowat, F. 1984. *Sea of slaughter*. Boston: Atlantic Monthly.

Cited in **General** Chapter as item 58.

* Mukherjee, A.K. 1982. *Endangered animals of India*. Calcutta: Zoological Survey of India.

Includes descriptions of 10 birds. See also **General** Chapter as item 59.

649. Mundy, P. 1982. The oxpeckers of Africa. *African Wildlife* 37(3): 110-116.

650. Munn, A. 1984. Black robin update. *Forest Bird* 15(2): 9-11.

The last two citations were not seen.

651. Munn, C.A. 1988. The real macaws. *Animal Kingdom* 91(5)Sept/Oct: 20-33.

Macaws are a tropical American bird. Eight of its 16 existing species are endangered and the other eight are declining dramatically. Threats are feather hunting, wild bird trade, and forest development. Article includes color photographs and drawings that aid identifying species. Diet and group behavior are described.

* National Wildlife Federation. 1988. Twentieth environmental quality index. *National Wildlife* 26(2) Feb-Mar: 38-46.

California condor, dusky seaside sparrow, and the bald eagle are among the birds mentioned. See item 62 in **General** Chapter.

652. Nawaz, M. 1982. Our vanishing birds. *Tigerpaper* 9(4): 20-21.

653. ______. 1984. Migratory cranes in Pakistan. *Tigerpaper* 11(4): 17-20.

The last two citations were not seen.

654. ______. 1984. Distribution and conservation of the houbara bustard (*Chlamydotis undulata* macqueen) in the province of Baluchistan (Pakistan). *Tigerpaper* 11(4): 5-9.

655. Newton, I. and E.D. Chancellor, eds. 1985. Conservation studies on raptors. Proceedings of the ICBP world conference on birds of prey held in Thessaloniki, Greece, April, 1982. *ICBP Technical Bulletin no. 5*.

Reports from several nations in the Mediterranean area describing bird counts and nesting sites. Comparisons with previous counts are noted. The classification "endangered" or "threatened" may vary among nations.

656. Niles, L.J., R.F. Harlow, and G.W. Woods. 1982. Home ranges of Red-cockaded woodpeckers in coastal South Carolina. *Auk* 99: 675-682.

657. Nilsson, G. 1981. The bird business. In *The coveted ones*. 2nd ed., 58-80. Washington, D.C.: Animal Welfare Institute.

* Nilsson, G. 1983. *The endangered species handbook.* Washington, D.C.: Animal Welfare Institute.

See item 63, **General** Chapter.

* Norden, A.W., D.C. Forester, and G.H. Fenwick. 1984. *Threatened and endangered plants and animals of Maryland.* Annapolis: Department of Natural Resources.

See item 64, **General** Chapter.

658. Norman, T. 1985. Hermit ibis in Turkey. *Oryx*. 19(1): 47.

* Norris, R. 1988. Data for diversity. *Nature Conservancy Magazine*. 38(1)Jan-Feb: 4-10.

Describes their method for collecting data, particularly biological surveys of birds in Central and South America. See item 65 in **General** Chapter.

* Nysewander, D.R., D.J. Forsell, P.A. Baird, D.J. Shields, G.W. Weiler, and J.H. Kogan. 1982. *Marine bird and mammal survey of the eastern Aleutian Islands, summers of 1980-1981*. Unpublished report. Anchorage: U.S. Fish and Wildlife Service.

Also noted as item 300 in **Mammals** Chapter.

659. Oakleaf, R. 1986. The home stretch. *Wyoming Wildlife*. 50(11): 8-16.

Not seen.

660. Odom, R.R. 1983. Georgia's red-cockaded woodpecker relocation experiment: a 1983 update. In *Red-cockaded woodpecker, Symposium II, Proceedings*. ed. D.A. Wood.

See item 794 for full citation.

661. Ogden, J. 1985. The California condor. *Audubon Wildlife Report 1985*. New York: National Audubon Society. pp. 388-399.

The *Audubon Wildlife Report* highlights topics of special interest; the condor appears in this issue.

661a. Ogden, J.C. 1983. The abundant, endangered flinthead. *Audubon* 85(1)Jan: 90,92,94-101.

One of three stork species native to America. Endangered status in Florida attributed to manipulation of water resources.

662. Olsen, P. 1985. Population studies of the peregrine in Australia. *ICBP Technical Publication No. 5*, 381-388.

663. Oren, D.C., and F.C. Novaes. 1986. Observations on the golden parakeet, *Aratinga guarouba*, in northern Brazil. *Golden Conservation* 36(4): 329-337.

664. Osborne, D.R., and A.T. Peterson. 1984. Decline of upland sandpiper (*Bartramia longicauda*) in Ohio: an endangered species. *Ohio Journal of Science* 84(1): 8-10.

665. Owen, M., G.L. Atkinson-Wiles, and D.B. Salmon. 1986. *Wildfowl in Great Britain*. 2nd ed. New York: Cambridge Univ. Press.

666. Padmanabhan, D., and C. Suderhersan. 1988. Mass destruction of *Phoenix loureirii* in south India. *Principes* 32(3): 118-123.

667. Palmer, M.J. 1985. Awak! The lively politics surrounding the sleepy spotted owl. *Pacific Discovery* 38(2): 24-25.

Conservation groups promoting the owl for listing on the federal list but loggers oppose.

668. Parker, K.E. 1987. Spirit of the northern waters. *Conservationist (New York)* 42(1)Jul/Aug: 16-21.

Considered a species of special concern in New York, loons have been taken for granted as summer residents of Adirondack lakes. Population is sparse but breeding pairs have a high reproductive rate. State monitors bird closely for effects of toxic chemicals, lake acidification and gear used by fishermen that can entangle loon.

669. Parslow, J.L.F., and M.J. Everett. 1981. Birds in need of special protection in Europe. *Nature Envir. Ser.* 24:1-154.

670. Pasquier, R., and C. Jones. 1982. The lost and lonely birds of Mauritius. *Natural History* 91(3): 39-43.

671. Pasquier, R.F., ed. 1982. *Conservation of new world parrots: Proceedings of the Second ICBP Parrot Working Group Meeting, St. Lucia, April 1980.* ICBP Technical Press.

672. Paton, P.W., J.M. Scott, and T.A. Burr. 1985. American coot and black-necked stilt on the island of Hawaii. *Western Birds* 16:175-181.

673. Perrins, C.M., and A.L.A. Middleton. 1985. *The encyclopedia of birds*. London and Sydney: George Allen and Unwin.

Includes species listed as threatened in the *Red Data Book*. Highlighted with a symbol.

674. Peterjohn, B.G., R.L. Hannikman, J.M. Hoffman, and E.J. Tramer. 1987. *Abundance and distribution of the birds of Ohio*. Biological Notes No. 19, Ohio Biological Survey. Columbus: Ohio State University.

Checklist of all species of native and naturalized birds documented to occur in Ohio. A note concerning their abundance or extirpation is included.

675. Petretti, A., and F. Petretti. 1981. A population of diurnal raptors in central Italy. *Gerfaut* 71(2): 143-156.

676. Philips, D., and H. Nash, eds. 1981. *The Condor question: captive or forever free?* San Francisco: Friends of the Earth.

Collection presents varying policy strategies of the attempts to preserve the California condor.

* Pilleri, G., and O. Pilleri. 1981. Some observations on mammals, birds and reptiles in the lower valley of the Indus and upper valley of the Brahmaputra. *Bollettino Societa Adriat. Sci. Nat.* 65(2): 133-166.

Cited in **Mammals** Chapter as item 317.

677. Pollock, H. 1986. The western bluebirds' struggle for survival. *Victorian Naturalist* 43(4): 1-4

678. Pratt, H.D., P.L. Bruner, and D.G. Berrett. 1987. *A field guide to the birds of Hawaii and the tropical Pacific*. Princeton: Princeton Univ. Press.

Provides special section on "status" denoting abundance, a table listing extinct birds, and a table of endangered or threatened breeding birds.

679. Quammen, D. 1982. Jeremy Bentham, the Pieta, and a precious few graylings. *Audubon* 84(3)May: 97-103.

Although the Michigan grayling became extinct in 1940, a remnant of Montana grayling endures.

680. Quammen, D. 1988. *The flight of the iguana*. New York: Delcorte Press.

Two birds extinct on Guam and three others nearly gone.

680a. Radtke, R.E., J. Byelich, and G. Irvine. 1983. Endangered bird of fire—the Kirtland's warbler. *Yearbook of Agriculture, 1983*. Washington, D.C.: U.S. Department of Agriculture.

681. Ramos, M.A. 1985. Endangered tropical birds in Mexico and northern Central America. *ICBP Technical Publication no. 4,* 305-318.

682. Rands, M. 1987. Action plan for birds. *Species* 8:28-29.

Lists ICBP priorities.

683. Rauzon, M.J. 1986. Spotlight on the kakapo. *Living Bird Quarterly* 5(1):30

Describes a rarely discussed bird.

* Raven, P.H. 1987. We're killing our world: preservation of biological diversity. *Vital Speeches* 53(May 15): 472-478.

Once there were 88 species of land birds in Hawaii, but now there are 28. See also **General** Chapter, item 72.

684. Redford, K.H. 1985. Emas National Park and the plight of the Brazilian cerrados. *Oryx* 19(4): 210-214.

685. Reed, J.R., J.L. Sinock, and J.P. Hailman. 1985. Light attraction in endangered procellariform birds: reduction by shielding upward radiation. *Auk* 102(2): 377-383.

686. Reed, S.M. 1981. New Zealand dotterel (Charadrius)- endangered species? *Notornis* 28(2): 129-132.

Argues that population of 1400 is endangered.

687. Reiger, G. 1987. Too many cooks. *Field and Stream* 92(Nov): 15-16.

Criticizes federal policy for being too manipulative and cites California Condor policy. See also **Mammals** Chapter, item 331.

688. Ripley, S.D. 1985. The laysan teal—recent history and the future? *Avicultural Magazine* 91(1-2): 76-78.

Restricted to a very small islet in the Pacific, the Laysan duck's population fluctuates wildly. Captive breeding suggested in the event a natural disaster eliminates the entire species during a period that the population is small.

689. Ripley, S.D., and B.M. Beehler. 1985. *Rails of the world, a compilation of new information, 1975-1983*. Smithsonian Contributions to Zoology. Washington, D.C.: Smithsonian Institution Press.

690. Robbins, J. 1988. Autumn of the eagles: has one of the world's greatest gathering of bald eagles vanished forever? *National Wildlife* 26(6)Oct-Nov: 34-40.

A site in Glacier National Park is declining as a migration halt puzzles scientists.

691. Robins, C.S., D. Brystak, and P.H. Geissler. 1986. *The breeding bird survey: its first 15 years, 1965-1979*. U.S. Fish and Wildlife Service, Resource Publication 157. Washington, D.C.

A joint U.S.-Canadian project that provides scientific and common names, and a paragraph for each species indicating population status.

692. Rogers, M.J. and the Rarities Committee. 1986. Report on rare birds in Great Britain. *British Birds* 79(11): 526-588.

Annual report of contributed sightings by birders identifying 147 species and the range of their sightings. Criteria for rarity is not stated.

* Roots, C. 1987. *Endangered species: Canada's disappearing wildlife.* Markham, Ont.: Fitzhenry & Whiteside.

Cited in **General** Chapter as item 75.

693. Ropper, P. 1988. Grand old ospreys. *Oceans* 21(1): 50-57.

Although not on any official "endangered" list, this author is concerned about their numbers.

694. Rounsevell, R.E., and J.C.Z. Woinarski. 1983. Status and conservation of the forty-spotted pardalote, *Pardalotus quadragintus. Australian Wildlife Research* 10(2): 343-349.

695. Rowe, B.E. 1985. The New Zealand kiwis (*Apteryx* sp.). *Avicultural Magazine* 91(1-2): 59-63.

Two of the three species remain and are described.

696. Rugis, S., and H. Schenk. 1981. Red list of Italian birds. *Avocetta: Bollettino della Associazione Ornitologica Italiana* 5(3): 133-141.

Not seen. Language is Italian but summaries are in English, French, and German.

697. Russell, R.P., Jr. 1983. The piping plover in the Great Lakes Region. *American Birds* 37(6)Nov-Dec: 951-955.

Gone from its range in the Great Lakes area, a viable population exists in Michigan. Author reports from surveys of north central states and Great Lake border states in U.S. and Canada and is not optimistic.

698. Ryden, H. 1983. Hatching high. *Geo* 5(Feb): 90-100.

The topic is the bald eagle.

699. Sailler, E., and B. McAllister. 1988. Pinnacle of glory. *International Wildlife* 18(2)Mar-Apr: 12-17.

Corsican ospreys threatened.

700. Sakai, H.F., C.J. Ralph, and C.D. Jenkins. 1986. Foraging ecology of the Hawaiian crow, an endangered generalist. *Condor* 88(2): 211-219.

* Sarkar, S.U. 1985. Endangered wildlife of the coastal zone of Bangladesh. *Coastal Zone* 1:1110-1120.

See **General** Chapter, item 77.

701. Sauey, R.T. 1986. The range, status and winter ecology of the Siberian crane

(*Grus leucogeranus*). *Dissertation Abstracts International* B 46(7): 2232-2233.

702. Savannah River National Environmental Research Park. 1981. *An annotated bibliography of the red-cockaded woodpecker*. Aiken, South Carolina: Savannah River Ecology Laboratory.

703. Savidge, J.A. 1987. Extinction of an island forest avifauna by an introduced snake. *Ecology* 68(3): 660-668.

The brown tree snake wreaked havoc on native birds.

704. Schenk, H. et al. 1985. The ecology of the Peregrine falcon in Italy: first results from five sample areas. *ICBP Technical Publication No. 5*, p. 367-380.

705. Schlorff, R.S., and P.H. Bloom. 1984. Importance of riparian systems to nesting Swainson's hawks in the central valley of California. In *California riparian systems: ecology, conservation, and productive management*, eds. R.E. Warner and K.M. Hendrix, 612-618. Berkeley, Los Angeles, and London: Univ. of California Press.

See **General** Chapter, item 103.

706. Schluter, P. 1988. The brown fish owl in Israel. *Israel Land and Nature* 13(4)Summer: 168-176.

Formerly considered rare, author writes that it is now extinct.

707. Schreiber, R.W. 1982. A brown study of the brown pelican. *Natural History* 91(1)Jan: 38-42.

708. Schreiber, R.W., and E.A. Schreiber. 1982. Essential habitat of the brown pelican in Florida. *Florida Field Naturalist* 10(1): 9-17.

709. Schulenberg, J.H., and M.B. Ptacek. 1984. Status of the interior least tern in Kansas. *American Birds* 38(6)Nov/Dec: 975-981.

Data from study determining its threatened status.

710. Schultz, M. 1987. *The hooded plover*. Royal Australasian Ornithological Report No. 35. (A RAOU Conservation Statement.)

Vulnerable throughout southern Australia.

711. Scott, D.A., and M.L. Brooke. 1985. The endangered avifauna of southeastern Brazil: a report of the BOU/WWF expeditions of 1980/81 and 1981/82. *ICBP Technical Bulletin No. 4*, 115-139.

An important cooperative venture between the Brazil Ornithological Union and the World Wildlife Fund.

712. Scott, J.M., C.B. Kepler, S.I. Peffer, and C. Van Riper, III. 1988. Conserva-

tion of Hawaii's vanishing avifauna. *Bioscience* 38(4)April: 238-254.

713. Scott, J.M., C.B. Kepler, P. Stine, H. Little, and K. Taketa. 1987. Protecting endangered forest birds in Hawaii: the development of a conservation strategy. In *Transactions of the 56th North American Wildlife and Natural Resources Conference,* ed. R.E. McCabe, 348-363.

714. Scott, J.M., S. Mountainspring, C. Van Riper, III, C.B. Kepler, J.D. Jacobi, T.A. Burr, and J.G. Giffin. 1984. Annual variation in the distribution, abundance, and habitat of the palila (*Loxioides bailleui*). *Auk* 101: 647-664.

715. Scowcroft, P.G. 1983. Tree cover changes in mamane forests grazed by sheep and cattle. *Pacific Science* 37(2): 109-119.

Argues that palila needs large trees for survival. Recent exclusion of feral sheep from Mauna Kea should permit growth of tree stock.

716. Seward, D. 1987. Europe's vanishing storks: breeding programs designed to reverse the birds' dramatic decline may actually be hurting the species' chances of survival. *International Wildlife* 17(3)May-June: 4-11.

717. Shapiro, A.E. 1983. Characteristics of red-cockaded woodpecker cavity trees and colony areas in southern Florida. *Florida Scientist* 46(2)Spring: 89-95. Orlando: Florida Academy of Sciences.

718. Shields, M.A., and T.W. Townsend. 1985. Nesting success of Ohio's endangered common tern. *Ohio Journal of Science*. 85(1): 45-50.

Descriptions of conditions on shores of Lake Erie.

719. Short, L.L. 1982. *Woodpeckers of the world.* Delaware Museum of Natural History, Monograph No. 4. Greenville, DE: Delaware Museum of Natural History.

A scientific treatment of all woodpecker species numbering about 200. While citing their endangered status is not primary to the book, rarity will be indicated in the paragraph entitled "Distribution and Habitat" within each species' description. The book lacks an index to those species considered endangered or rare.

720. ______. 1985. Last chance for the ivorybill. *Natural History*. 94(8)Aug 66-68.

721. ______, and J.F.M. Horne. 1986. The ivorybill still lives. *Natural History* 95(7)Jul: 26,28.

An account of sighting a pair of ivorybills in Cuba.

722. ______. 1987. I saw it. *International Wildlife* 17(2)Mar-Apr: 22-23.

In the rugged mountains of eastern Cuba scientists spot the ivory-billed woodpecker, a bird once thought extinct.

723. Sibson, R.N. 1982. *Birds at risk: rare or endangered species of New Zealand*. Illustrations by B. Howard. Wellington, N.Z.: A.H. and A.W. Reed.

Excellent description of 27 species accompanied by colored plates.

724. Sidle, J.G., and D.B. Bowman. 1988. Habitat protection under the Endangered Species Act. *Conservation Biology* 2(1)Mar: 116-118.

Paper describes how the Act preserves habitat critical for endemic species. Cites the Hawaiian palila as a species dependent on specific trees and forests. Palila is a honeycreeper endemic to Mauna Kea. Least tern and piping plover are other examples.

725. Silva, T. 1985. Decline of the greater Patagonian conure (*Cyanoliseus patagonus* byroni). *Avicultural Magazine* 91(1-2): 64.

This is a one page article but the species is endangered and not mentioned elsewhere.

726. ______. 1988. *A monograph of endangered parrots*. Illusrated by Gracia Bennish. Ontario: Silvio Mattacchione & Co. 1793 Rosebank Road, N. Pickering, Ontario, Canada.

A limited edition due for delivery in 1989 but started in production 1987.

729. Silvius, M., W. Verheugt, and J. Iscandar. 1985. Indonesian milky storks and Asiatic dowitchers in Sumatra. *Interwader* 5:1-3.

730. Simons, T., S.K. Sherrod, M.W. Collopy, and M.A. Jenkins. 1988. Restoring the bald eagle. *American Scientist* 76(3)May-June: 253-260.

Describes a variety of strategies, including hacking, to increase the population of eagles. Discussion reveals the causes of the decline and the complexity of planning recovery. Good bibliography.

731. Simons, T.R. 1985. Biology and behavior of the endangered Hawaiian dark-rumped petrel. *Condor* 87(2): 229-245.

732. ______. 1983. *Biology and conservation of the endangered Hawaiian dark-rumped petrel (Pterodroma phaeopygia sandwichensis)*. Seattle: National Park Service, Cooperative Park Studies Unit, College of Forestry Resources, University of Washington.

See also *Dissertation Abstracts International* 44(04-B): 966, for a review of the author's Ph.D. thesis (1983, University of Washington) with the same title.

* Skoog, P.J. 1982. *Highways and endangered wildlife in Florida: impacts and recommendations*. (Final report). Sponsor: Federal Highway Administration, Florida Game and Freshwater Fish Commission. Tallahassee, Florida.

See **General** Chapter item 82.

733. Skutch, A.F. 1982. Resplendent myth. *Audubon* 84(5)Sept: 74-84.

Description and photos of the quetzel, a bird linked to pre-Colombian culture.

734. Snyder, H. 1986. Changing times for the California condor. *Zoonooz* 59(5)May: 4-9.

Describes death of last female of the wild breeding pairs from eating a carcass that contained a lead bullet. Controversy about where to release birds raised in captivity. Estimate that 5 wild birds remain.

735. Snyder, N.F.R., J. Wiley, and C. Kepler. 1987. *The parrots of Luquillo: natural history and conservation of the Puerto Rican parrot.* Los Angeles: Western Foundation of Vertebrate Zoology.

736. Sollen, R. 1982. Comeback of the pelicans. *Defenders* 57(1)Feb: 19-25.

Pelican decline began to be noticed in 1969. Research into causes discovered reproductive failure due to thin egg shells. Pesticides in the environment were the culprit and restoration efforts led to significant clean up of marine habitats. Author does not believe pelican recovery adequate to justify delisting.

737. Sophasan, S., and R.N.G. Dobias. 1984. The fate of the "Princess Bird", or white-eyed river martin. *Natural History Bulletin of Siam Society* 32(1):1-10.

738. Soule, M. et al. 1988. Reconstructed dynamics of rapid extinctions of chaparral-requiring birds in urban habitat islands. *Conservation Biology* 2(1): 75-92.

739. Spencer, R. and Rare Breeding Birds Panel (Comps). 1988. Rare breeding birds in the United Kingdom in 1985. *British Birds* 81:99-125.

740. Spina, F., H. Schenk, and B. Massa. 1985. Status and conservation of Eleanora's falcon in Italy. *ICBP Technical Publication. No. 5,* 143-146.

741. Spitzer, P. 1981. The lily of birds. *Animal Kingdom* 84(5)Oct-Nov: 24-30.

This elegant crane, which often winters in India, is one of the world's most endangered. Study of feeding habits and ecology recommended to prevent loss to competing crane populations.

742. Stalcaster. M.V. 1987. *The bald eagle.* New York: Universe Books/St. Martin's Press.

An in-depth treatise on a bird that is universally recognized as endangered.

* Steadman, D.W., and S. Zousmer. 1988. *Galapagos; discovery on Darwin's islands.* Drawings by L.M. Steadman. Washington, D.C.: Smithsonian Institution Press.

See **General** Chapter item 85.

743. Steinhart, P. 1987. Empty the skies. *Audubon* 89(6)Nov: 71-95.

Argues that the decline of birds is due to the loss of wetlands.

744. ______. 1983-84. Trouble in the tropics. *National Wildlife* 22(Dec-Jan): 16,18,20.

Songbirds threatened by elimination of Central and South American forests as winter habitats.

745. _______. 1986. Synthetic species. *Audubon* 88(5)Sept: 8-11.

* Stermer, D. 1981. *Vanishing creatures, a series of portraits.* Introduction by Robert Redford. Berkeley, Calif.: Lancaster-Miller Publishers.

See item 86 in **General** Chapter.

746. Stone, C.P., R.L. Walker, J.M. Scott, and P.C. Banko. 1983. Hawaiian goose management and research—where do we go from here? *Elepaio* 44:11-15.

747. Storer, R.W. 1982. The hooded grebe on laguna de los Escarchados: ecology and behavior. *Living Bird* 19:51-67.

Not seen.

748. Strange, I.J. 1986. The Falklands' Johnny Rook. *Natural History* 95(2)Feb: 54-60.

Once thought "exceedingly common" by Darwin, its breeding areas have diminished. A complicated food pattern is described.

749. Stuart, S.N. 1985. Rare forest birds and their conservation in eastern Africa. *ICBP Technical Publication No. 4*, 187-196.

750. Stuart, S.N. 1984. ICBP Cameroon montane forest survey. *Malimbus* 6(1-2): 5-7.

A summary of the ICBP study. See item 88 in **General** Chapter.

* Stuart, S.N., ed. 1986. *Conservation of Cameroon montane forests.* Report of the ICBP Cameroon Montane Forest Survey, Nov. 1983-April 1984. Cambridge, England: International Council for Bird Preservation.

See item 88, **General** Chapter.

751. Sullivan, J. 1986. Lake Renwick: unlikely haven for the endangered. *Bulletin, Field Museum of Natural History* 51(5/May): 10-17.

Three species are officially endangered in Illinois.

752. Sullivan, S. 1982. Endangered...extinct? *Bird Life* Jan-Feb: 36-39.

Not seen

753. Swengel, S. 1985. Genetic and demographic management of captive red-crowned cranes (*Grus japonensis*). *Avicultural Magazine* 91(1-2): 41-47.

An endangered species that has become well established in captivity. Author describes problems of genetic management of maintaining a captive population and inbreeding.

754. Swenson, J.E., K.L. Alt, and R.L. Eng. 1986. Ecology of bald eagles in the greater Yellowstone ecosystem. *Wildlife Monographs No. 95.*

The results of a long term (1972-1982) study which is rare among studies of eagles.

755. Swenson, J.E., C.A. Simmons, and C.D. Eustace. 1987. Decrease of sage grouse, *Centrocerus urophasianus*, after ploughing of sage brush steppe. *Biological Conservation* 41(2): 125-132.

Defends need for unploughed fields of sagebrush (artemesia) for lekking males as wintering areas. Population declines shown when fields are ploughed.

756. Tacha, T.C. 1988. Social organization of sandhill cranes from mid-continental North America. *Wildlife Monographs No. 99* (Oct.)

The Mississippi sandhill crane is on the U.S. Endangered Species List.

757. Tangley, L. 1986. The urban ecologist. *Bioscience* 36(2)Feb: 68-71.

Describes project to establish viable populations of Peregrine falcons in high bridges and buildings of cities.

758. Tate, J., Jr. 1986. The blue list for 1986. *American Birds* 40(2): 227-236.

Annual list of threatened birds as determined by the AOU. "Blue Lists" for each year can be found in each volume of *American Birds* since 1971.

759. _______. 1981. The Blue List for 1981. *American Birds* 35(1)Jan: 3-10.

Summarizes the first 10 years of the annual "early warning list" of declining, threatened, or vulnerable species developed by the AOU. Species are nominated by sighters and bird counts and are NOT necessarily on the U.S. Endangered Species List.

760. Taylor, R.H. 1985. Status, habits and conservation of cyanoramphus parakeets in the New Zealand region. *ICBP Technical Publication No. 3*, 195-211.

761. Teixeira, D.M., and D.W. Snow. 1982. Notes on the nesting of the red-billed curassow *Crax Blumenbachii*. *Bulletin, British Ornithological Club* 102(2): 83-84.

A Brazilian bird.

762. Telfair, R.C., and D.A. Swepston. 1987. *Analysis of banding and marking*

nestling anhingas, olivaceous cormorants, roseate spoonbills, ibises, bitterns, herons and egrets in Texas (1923-1983). Austin: Texas Parks and Wildlife Department, Nongame and Endangered Species Program.

763. Temple, S.A. 1987. The problem of avian extinctions *Current Ornithology* 3:453-485.

764. Terrasse, M. 1982. Concrete measures. *Naturopa* 40: 18-20.

The effect of building and development on bird habitat.

765. Tidwell, M. 1985. Warheads and woodstorks. *Sierra* 70(1):14. Habitat destruction of the endangered wood stork.

* Tikader, B.K. 1983. *Threatened animals of India.* Calcutta: Zoological Survey of India. With a foreword by T.N. Koshoo.

See item 89, **General** Chapter.

766. Todd, D.M. 1984. The release of pink pigeons *Columba (nesoenas) mayeri* at Pamplemousses, Mauritius—a progress report. *Dodo* No. 21:43-57.

767. Todd, F.S. 1981-1982. A gathering of penguins. *Animal Kingdom* 84(6)Dec-Jan: 16-23.

Author claims population should be considered endangered as it now numbers fewer than 10,000. Describes San Diego's Sea World's efforts to establish breeding populations to maintain the species as unusually difficult.

768. Tomkins, R.J. 1985. Breeding success and mortality of dark rumped petrels in the Galapagos, and control of their predators. *ICBP Technical Publication No. 3,* p. 159-175.

769. Tomlinson, D. 1982. A question of survival: endangered birds. *Country Life* 171(4409): 398-400.

770. Toone, C.K. 1986. The Condor's last stand. *Zoonooz* 59(1)Jan: 9-11.

Biologists estimate only six birds left in the wild. Overview of causes of decline and several breeding strategies.

771. U.S. Department of the Interior, Bureau of Land Management. 1982 (June). *Twin Falls livestock grazing, Twin Falls County, Idaho.* Environmental Impact Statement. 2 vol. Burley, Idaho: Department of the Interior, Bureau of Land Management.

Long-billed curlews and burrowing owls are among the birds of concern. See also item 379 in **Mammal** Chapter.

772. U.S. Department of the Army (Corps of Engineers) and Department of the Interior, Fish and Wildlife Service. 1982. *Bottomland hardwood preservation program, Reeves Bottom (Hornersville Swamp), Dunklin County, Missouri.* Environmental Impact Statement. Jackson, Mississippi: U.S.

Department of the Interior, Fish and Wildlife Service.

Bald eagles, Peregrine falcons, and Bachmans' warbler are among the birds of concern. See also item 380 in **Mammals** Chapter.

773. U.S. Fish and Wildlife Service. 1985. *Migratory bird hunting. Federal Register* 50(88) May 7, p.19248.

Legislation and regulations concerning hunting and endangered birds.

774. _______. 1985. *Red-cockaded woodpecker recovery plan.* Atlanta: U.S. Fish and Wildlife Service.

Conservation planning.

775. Urban, E.K., C.H. Fry, and S. Keith, eds. 1982 and 1986. *The birds of Africa.* 2 vols. London: Academic Press.

Arranged by systematic classification, each species receives a detailed, traditional description. Within the "range and status" subsection, the abundance is noted as common, extinct, rare, very rare, scarce, not threatened, threatened, or uncommon according to the authors' opinion.

776. Vernon, A. 1987. No room for a hermit. *International Wildlife* 17(6)Nov-Dec: 23-24.

Maori people call it the "hoiho" and it is considered the world's rarest penguin.

777. Walker, R.W. 1982. The wattled crane (an endangered species). *Redwing.* 1982: 6-8.

778. Walkinshaw, L.H. 1983. Kirtland's warbler: the natural history of an endangered species. *Cranbrook Institute of Science Bulletin 58.* Bloomfield Hills, Michigan: Cranbrook Institute.

Article describes the species and the factors of its decline. A fine bibliography is included.

* Wallace, D.R. 1987. *Life in the balance.* New York, San Diego, London: Harcourt Brace Jovanovich.

A popular description of most well-known endangered birds: condor, wood stork, ivory and red-cockaded woodpeckers, peregrine falcons, and eagles. Beautiful photographs. Discussion organized around ecosystems.

See also **General** Chapter, item 101.

779. Wallace, M.P. 1985. Ecological studies of Andean condors in Peru. *Dissertation Abstracts International* 46(09-B): 2978. Ph.D., University of Wisconsin-Madison.

Purpose of the study is to observe behavior of Andean condor to identify strategies that may be used in preserving the California condor.

780. Wallis, D. 1986. *The rare breeds handbook.* New York: Blandford; Dorset, UK: Poole.

Not seen.

781. Walters, M. 1985. On the status of *Psittacula intermedia* (Rothschild). *Journal of the Bombay Natural History Society* 82(1)April: 197-199.

Controversial taxonomy of a small bird group not seen in many years suggesting extinction or highly endangered. Includes parakeets.

782. Watson, J. 1984. Land birds: endangered species on the granitic Seychelles. *Monographiae Biologia* 55:513-527.

783. Weathers, W.W., and C. van Riper III. 1982. Temperature regulation in two endangered Hawaiian honeycreepers: the palila (*Psittirostra bailleui*) and the Laysan finch (*Psittirostra cantans*). *Auk* 99:667-674.

784. Weaver, C.M. 1982. Breeding habitats and status of the golden-shouldered parrot, *Psephotus chrysopterygius*, in Queensland. *Emu* 82(1): 2-6.

785. Welch, G. and H. Welch. 1985. Observations on the endemic Djibouti francolin (*Francolinus ochropectus*) with emphasis on potential threats to the population. *Journal of the World Pheasant Association* 10:65-71.

* Westing, A.H., and C.E. Westing. 1981. Endangered species and habitats of Viet Nam *Environmental Conservation* 8(1): 59-62.

See item 104 in **General** Chapter.

786. Wexler, M. 1986. Can we save our only parrot? *National Wildlife* 24(Aug-Sept): 4-8.

Puerto Rican parrot species nearing extinction.

787. Wheeler, W. 1982. Season of the eagle. *National Parks* 56(11-12): 22-24.

* White, C.P. 1982. *Endangered and threatened wildlife of the Chesapeake Bay Region: Delaware, Maryland, and Virginia.* Centerville, Md.: Tidewater Publishers.

Cited as item 105 in **General** Chapter.

788. Whitlock, R. 1981. *Birds at risk: a comprehensive world survey of threatened species.* Atlantic Highland, N.J.: Humanities Press; Bradford-on-Avon, England: Moonraker Press.

789. Whitman, P.L. 1988. *Biology and conservation of the endangered interior least tern: a literature review.* U.S. Fish and Wildlife Service Biological Report 88(3). Washington,D.C.: Department of the Interior, Division of Endangered Species.

Text of 10 pages reviews taxonomy, description, range, life history, and present status of bird. Nine pages listing 161 citations of interior least tern literature. List is valuable for inclusion of state, regional, and federal publications, surveys, reports and journal literature.

790. Wieloch, M. 1984. Numbers and distribution of the mute swan *Cygnus olor* in Poland against the situation of this species in Europe. *Acta Ornithologia Warszaw* (English) 20(2): 187-240.

791. Willis, E.O., and Y. Oniki. 1982. Behavior of fringe-backed fire-eyes, *Pyriglena atra,* (Formicariidae): a test case for taxonomy versus conervation. *Revista Brasilian Biologia* 42(1): 213-223.

792. ______. 1981. Notes on the slender antbird. *The Wilson Bulletin* 93(1)Jan: 103-107.

A brief note in which authors argue for IUCN inclusion of this bird.

* Wilson, J.H. comp. and ed. 1984. *Rare and endangered species of Missouri.* Jefferson City, MO: Missouri Department of Conservation.

Many bird species noted. See **General** Chapter item 108.

793. Woinarski, J.C.Z., and B.J. Wykes. 1983. Decline and extinction of the helmeted honeyeater at Cardinia Creek. *Biological Conservation* 27(1): 7-21.

794. Wood, D.A., ed. 1983. *Red-cockaded woodpecker symposium II, proceedings.* Tallahassee: Florida Game and Freshwater Fish Commission. Washington, D.C.: U.S. Fish and Wildlife Service and the U.S. Forest Service.

Individual papers from these *Proceedings* have been cited throughout this chapter.

795. Wood, G.W., L.J. Niles, R.M. Hendrick, J.R. Davis, and T.L. Grimes. 1985. Compatibility of even-aged timber management and Red-cockaded woodpecker conservation. *Wildlife Society Bulletin* 13:5-17.

Plan to maintain adequate pine forests for preservation of this species.

796. Wood, L. 1983. Plight of the jackass penguin. *Sea Frontiers* 29(6): 350-358.

797. Woods, C.A., 1987. The threatened and endangered birds of Haiti: lost horizons and new hopes. *Proceedings of the Jean Delacour/IFCB symposium on breeding birds in captivity,* 2nd, International Foundation for the Conservation of Birds. San Diego, California. pp. 385-429.

Includes a loose-leaf, illustrated publication that identifies rare birds being reproduced in captivity. Also cited as item 568.

* Woolfenden, G.E. 1983. *Rare, threatened, and endangered vertebrates of southwest Florida and potential OCS (Outer Continental Shelf) activity impacts.* Washington, D.C.: U.S. Fish and Wildlife Service, Office of

Biological Services, and Bureau of Land Management, Branch of Offshore Studies.

See item 112 in **General** Chapter.

798. Wyckoff, A.M. 1986. A relict population of chestnut collared longspurs in western Minnesota. *Loon* 58(1):3-11.

* Yaffee, S.L. 1982. *Prohibitive policy: implementing the Federal Endangered Species Act.* Cambridge, MA: MIT Press

See item 113 in **General** Chapter.

799. Yates, S. 1984. On the cutting edge of extinction. *Audubon* 86(4)July: 62-86.

What's left of the nene.

800. Yinzeng, L. 1982. Search for the crested ibis. *New Scientist* 94(1301): 171-172.

801. Zeleny, L. 1983. So keep on looking for the blue bird. *Living Bird Quarterly* 2(2)Spring: 18-22.

Reasons for decline of eastern bluebird and recommendations for revival of population. Eastern bluebird has declined 90% in last 50 years. Severe winters, use of pesticides in orchards, decline of small farms (their favorite habitat), cavity nesting, habitat supply going because of removal of dead trees. Suggests planting for gardens.

802. Zembal, R. et al. 1985. Intermarsh movements by light-footed clapper rails indicated in part through regular censusing. *California Fish and Game* 71(3): 164-171.

803. Zembal, R., and B.W. Massey. 1985. Distribution of the light-footed clapper rail in California, 1980-1984. *American Birds* 39(2) Summer: 135-137.

Identifies California nest sites included in census. Small bibliography includes authors' previous publications.

804. Zhi-Yen, Z. 1986. Gem of the orient. *Living Bird Quarterly* 5(2): 29-30.

805. Zwank, P.J., J.P. Geaghan, and D.A. Dewhurst. 1988. Foraging differences between native and released Mississippi sandhill cranes, implications for conservation. *Conservation Biology* 2(4): 386-390.

AMPHIBIANS AND REPTILES

* Aiken, S.R., and C.H. Leigh. 1985. On the declining fauna of peninsular Malaysia in the postcolonial period. *Ambio* 14(1): 15-23.

See item 1 in **General** Chapter.

806. Alderton, D. 1988. *Turtles and tortoises of the world.* Photographs by Tony Tilford. New York: Facts on File.

Although very comprehensive in attention to all existing turtles and tortoises, those that are considered endangered and rare are noted. Wonderful natural history descriptions.

* Anadu, P.A. 1987. Progress in the conservation of Nigeria's wildlife. *Biological Conservation.* 41(4): 237-251.

Cited in **General** Chapter, item 3.

807. Arnold, P. 1983. Come back Kemp's Ridley. *Science 83* 4(8): 68-71.

808. Auffenberg, W. 1988. *Gray's monitor lizard.* Gainesville: Univ. Presses of Florida.

One of the varanic lizards, Gray's monitor differs from others in the species. A biological and ecological presentation accompanied by conservation recommendations. Timber harvesting and illegal hunting put this lizard at risk.

809. _______. 1981. *The behavioral ecology of the Komodo monitor.* Gainesville: Univ. Presses of Florida.

The "dragon" of Indonesia.

810. ______, and R. Franz. 1982. The status and distribution of the gopher tortoise (*Gopherus polyphemus*). *Wildlife Research Report* 12. U.S. Fish and Wildlife Service, pp. 95-126.

811. Bacon, J.P., and R.P. Reynolds. 1982. Current status of the Galapagos tortoise (*Geochelone elephantopus*) in Ecuador. *AAZPA Ann. Conference Proceedings of 1982.* pp. 414-423.

812. Bartlett, R.D. 1988. *In search of reptiles and amphibians.* Leiden and New York: E.J. Brill.

813. Beebee, T.J.C. 1983. *The natterjack toad.* Oxford: Oxford Univ. Press.

Protected under the Conservation of Wild Creatures and Wild Plants Act of 1975 because it is one of the most endangered species in Great Britain. Everything in 156 pages that you could possibly want to know about the natterjack toad.

* Bendiner, R. 1981. *The fall of the wild, the rise of the zoo*. New York: E.P. Dutton.

Cited in **General** Chapter item 10.

* Berger, T.J., A.M. Neuner, and S.R. Edwards. 1981. *Directory of federally controlled species*. Lawrence, Kansas: Association of Systematics Collections.

Cited in **General** Chapter item 11.

814. Bickham, J.W., M.D. Springer, and B.J. Gallaway. 1984. Distribution survey of the yellow mud turtle (*Kinosternon flavescens*) in Iowa, Illinois, and Missouri: a proposed endangered species. *Southwestern Nature* 29(1): 123-132.

815. Bjorndal, K.J., ed. 1982. *Biology and conservation of sea turtles; proceedings of the World Conference on Sea Turtle Conservation, Washington, D.C., 26-30 November 1979*. Washington, D.C.: Smithsonian Institution Press in cooperation with World Wildlife Fund.

* Boyd, J.M. 1984. Nature conservation. *Proceedings of the Royal Society of Edinburgh*. B-Biological Sciences 84: 295-335.

Cited in **General** Chapter as item 12. Natterjack toad is considered.

816. Branch, W.R. 1983. The dwarf angolan python, *Python anchietae*: an endangered species. *Litteratura Serpentium* 3(4): 121-125.

817. Breisch, A.R. 1984. Just hanging in there: the eastern massasauga, in danger of extinction. *Conservationist* (New York) 39(3): 35-38.

818. Bruton, M.N. 1982. Uncommon and rare reptiles of Maputaland. *African Wildlife* 36(4-5): 184-185.

Not seen.

819. Buitrago, J. 1985. Will the Caribbean Hawksbill turtle survive? *Sea Frontiers* 31(4): 219-226.

820. Burbidge, A.A. 1981. The ecology of the western swamp tortoise, *Pseudemydura umbrina*, (Testudines: Chelidae). *Australian Wildlife Research* 8(1): 203-223.

821. Burghardt, G. M., and A.S. Rand, eds. 1982. *Iguanas of the world: their behavior, ecology, and conservation. Noyes Series in Animal Behavior, Ecology, Conservation and Management*. Park Ridge, N.J.: Noyes.

Chapters by prominent researchers are included for a comprehensive examination of the current biology of this lizard, but it is most persuasive in describing the many species in this unique taxon that are on the verge of extinction.

822. Bury, R.B., ed. 1982. North American tortoises: conservation and ecology. *Wildlife Research Report No. 12*. U.S. Fish and Wildlife Service.

* Byrne, G. 1988. Strengthened endangered species act passes. *Science* 242(4876)Oct 14: 190.

  Kemp's ridley sea turtle is mentioned. Cited in **General** Chapter as item 14.

* Cadieux, C. 1981. *These are the endangered.* Washington, D.C.: Stone Wall Press; Brattleboro, Vt.: Distributed by Stephen Greene Press.

  Cited in **General** Chapter item 15.

823. Carr, A.F. 1986. Rips, fads, and little loggerheads. *Bioscience* 36(2)Feb: 92-100.

  Explores the unknowns of the reproduction journeys of the green and loggerhead turtles from the Azores to the Caribbean in effort to improve conservation strategy.

824. ______. 1984. *So excellent a fish: a natural history of sea turtles*. New York: Scribner's.

  Originally published New York: Doubleday, 1967. This edition has a new epilogue and photographs. Also published in paperback, 1986, Univ. of Texas Press.

825. Christiansen, J.L. et al. 1985. Aspects of the natural history of the yellow mud turtle, *Kinosternon flavescens* (Kinosternidae) in Iowa: a proposed endangered species. *Southwestern Nature*. 30(3): 413-425.

* Coffin, B., and L. Pfannmuller, eds. 1988. *Minnesota's endangered flora and fauna*. Minneapolis: Univ. of Minnesota Press for the Natural Heritage and Nongame Wildlife Programs of the Division of Fish and Wildlife, Minnesota Department of Natural Resources.

  Cited in **General** Chapter item 17.

826. Conner, D.K., and R. O'Dell. 1988. The tightening net of marine plastics pollution: strategies for intervention. *Environment* (Washington, D.C.) 30(1)Jan-Feb: 16-20, 33-36.

  Describes role of plastic refuse in killing marine life, e.g. green turtles and manatees.

* Convention on International Trade in Endangered Species (CITES). 1981- *Identification manual.* ed., P. Dollinger. 5 v. (loose-leaf) v. 3. Reptilia. Amphibia. Pisces.

Cited in **General** Chapter item 19.

827. Corke, D. 1987. Reptile conservation on the Maria Islands (St. Lucia, West Indies). *Biological Conservation* 40(4): 263-279.

A check-list of reptile fauna of St. Lucia.

* Cypher, E.A., R.H. Yahner, G.I. Storm, and B.L. Cypher. 1986. Flora and fauna survey in a proposed recreational area of Valley Forge National Historical Park. *Proceedings of the Pennsylvania Academy of Sciences*. 60(1): 47-50.

Cited in **General** Chapter item 23.

* DiSilvestro, R. L., and A.S. Eno, eds. 1986. *Audubon wildlife report 1986*. Audubon Wildlife Report (Annual). New York: National Audubon Society.

Cited in **General** Chapter item 26.

828. Dixon, J.R. 1987. *Amphibians and reptiles of Texas: with keys, taxonomic synopses, bibliography and distributional maps*. W.L. Moody, Jr. Natural History Series no. 8. College Station: Texas A & M Univ. Press.

A scientific work giving anatomical keys for identification of species and references to original descriptive publications. Although lacking narrative descriptions about population status or natural history a lengthy bibliography refers to works of that type. Reader needs other source to identify rarity or endangerment.

829. ______, and P. Soini. 1986. *The reptiles of the Upper Amazon Basin, Iquitos Region, Peru*. Part 1. Lizards and amphibians. Part 2. Crocodilians, turtles and snakes. ed. Mary Garity. Milwaukee: Milwaukee Public Museum. These two parts originally published separately in 1975 and 1977. Current volume is update and combination of the two.

A work for biological systematists, the taxonomy of the eastern indigo snake, on the U.S. Endangered List, is presented in addition to its morphology and bibliographic references.

* Doughty, R.W. 1983. *Wildlife and man in Texas*. College Station: Texas A & M University Press.

Cited in **Mammals** Chapter item 180.

* Durrell, G.M. 1983. *Ark on the move*. New York: Coward-McCann.

Cited in **Mammals** Chapter item 181.

830. Ehmann, H. 1983. The natural history and conservation status of the Adelaide Pigmy Bluetongue Lizard, *Tiliqua adelaidensis*. *Herpetofauna*. 14(1-2): 61-76.

Considered rare by the author.

832. Endemic Species Committee, Colorado River Wildlife Council. 1982. *Endemic amphibians and reptiles of the Colorado River System: a status report.* [Denver]: Colorado River Wildlife Council.

This 44 page booklet is useful as an identification list of 126 herpetofauna found in the defined area. The status of each species is categorized as common, uncommon, rare (seldom found in habitat), undetermined, peripheral (occuring at margins of its natural habitat), depleted (numbers greatly reduced), threatened (may likely become endangered), and endangered (may likely become extinct).

* Franklin, K.E. 1987. Endangered species: where to from here? *American Forests* 93(11-12)Nov-Dec: 57-60, 74-76.

Cited in **General** Chapter item 33.

* Fritts, T.H. et al. 1983. *Turtles, birds, and mammals in the northern Gulf of Mexico and nearby Atlantic waters; an overview based on aerial surveys of OCS* [outer continental shelf] *areas, with emphasis on oil and gas effects.* FWS/OBS-82/65. Washington, D.C.: U.S. Fish and Wildlife Service, Minerals Management Service.

Cited in **General** Chapter item 34.

832. Gallaway, B.J., J.W. Bickham, and M.D. Springer. 1985. A controversy surrounding an endangered species listing: the case of the Illinois mud turtle, another perspective. *Smithsonian Herpetology Information Service.* 64:1-17.

See article's bibliography for other writings on this subject.

* Genoways, H., and F.J. Brenner, eds. 1985. *Species of special concern in Pennsylvania.* Compiled by the Pennsylvania Biological Survey. *Special Publication of the Carnegie Museum of Natural History* No. 11. Pittsburgh: Carnegie Museum of Natural History.

Cited in **General** Chapter item 35.

833. Gibbons, J. 1984. Iguanas of the South Pacific. *Oryx* 18(2): 82-91.

834. Ginsberg, J. 1981. The status of sea turtles in Tarutao National Park, Satun, Thailand. *Tigerpaper.* 8(2): 27-29.

835. Glastra, R. 1983. Notes on a population of Caiman crocodilus crocodilus depleted by hide hunting in Surinam. *Biological Conservation* 26(2): 149-162.

The larger crocodilians have become virtually extinct in Latin America so that concern is now centered on the smaller species. The IUCN listed the spectacled caiman as endangered in 1979. Paper is an inventory of the species' number in certain areas.

836. Grigg, G., R. Shine, H. Elmann, eds. 1985. *Biology of Australasian frogs*

and reptiles. Chipping Norton, Australia: Surrey Beatty Ltd. in cooperation with the Royal Zoological Society of New South Wales.

A collection of papers from the Australasian Herpetological Conference, August 1984. Nine articles concern rare and endangered species.

837. Groombridge, B. 1981. *World checklist of endangered amphibians and reptiles*. Rev., 2nd ed. London: Wildlife Advisory Branch, Nature Conservancy Council.

838. ______. comp. Assisted by L. Wright, with the help of the Species Survival Commission of the IUCN. 1982. *Testudines, crocodylia, and rhynchocephalia. IUCN Amphibia-reptilia Red Data Book,* Part I. Gland, Switzerland: IUCN, and Cambridge, England: IUCN Conservation Monitoring Ctr.

As usual, a very authoritative work from this organization. Bibliography, references and index. Supercedes *World checklist of endangered amphibians and reptiles*, 2nd ed., Nature Conservancy Council, 1981.

* Groves, R.H., and W.D.L. Ride, eds. 1982. *Species at risk: research in Australia*. Proceedings of a symposium on the biology of rare and endangered species in Australia. Canberra and New York: Springer-Verlag.

Cited in the **General** chapter item 37.

839. Halliday, T., and K. Adler, eds. 1986. *Encyclopedia of reptiles and amphibians*. New York: Facts on File.

Endangered and rare species are so noted.

840. Honegger, R.E. 1981. *Threatened amphibians and reptiles in Europe*. (Supplementary volume of *Handbuch der reptilien und amphibien Europas*) European Committee for Conservation of Nature and Natural Resources Council of Europe. Wiesbaden: Akademische Verlagsgesellschaft.

A welcome addition for the English reader as the original volumes are in German. Revised edition of the 1st attempt to establish a list of threatened European amphibia and reptile taxa. Threatened amphibia and reptilia are identified, range and status given, and causes for decline are examined in a scientific manner.

* Howard, R., A. Perry, P. Herring, and J. Scott (Study Manager.) 1984. *Endangered and threatened species on U.S. Air Force installations*. Fish and Wildlife Service/Office of Biological Services (FWS/OBS) 84/110. Washington, D.C.: U.S. Fish and Wildlife Service, Department of the Interior.

Cited in **General** Chapter item 42.

841. Huang, Z., H. Lin, and S. Zhang. 1986. Analysis of the landset remote sensing images of the types of habitats of yangtse alligators. *Chinese Journal of Oceanology and Limnology*. 4(4): 360-371. The Chinese alligator is on the U.S. Endangered list.

842. Hughes, G. 1982. The beacon fades at Bhanga Nek. *African Wildlife* 36(4-5): 131-134.

Not seen.

843. Hutton, J.M. 1987. Growth and feeding ecology of the Nile crocodile (*Crocodylus niloticus*) at Ngezi, Zimbabwe. *Journal of Animal Ecology*. 56(1): 25-38.

Not seen.

* International Convention Advisory Commission. 1981. *Convention on international trade in endangered species of wild fauna and flora*. Washington, DC:

CITES cited in **General** Chapter item 43.

* IUCN Conservation monitoring centre. 1986. *1986 IUCN red list of threatened animals*. Prepared by the IUCN Conservation Monitoring Centre, Cambridge, U.K.

Cited in **General** Chapter item 44.

* _______. 1988. *1988 IUCN red list of threatened animals*. Cambridge, U.K., and Gland, Switzerland.

Cited in **General** Chapter item 45.

844. Jeffrey, R.C.V. 1982. A review of the status of crocodiles in Zambia. *Black Lechwe* 2(N.S.): 20-21.

A difficult journal to find. See Appendix A for address of publisher, Wildlife Conservation Society of Zambia.

845. Johns, A.D. 1987. Continuing problems for Amazon River turtles. *Oryx* 6(1): 25-28.

846. Khan, M.A.R. 1982. Chelonians of Bangladesh and their conservation. *Journal of the Bombay Natural History Society* 79(1): 110-115.

847. Kiew, B.H. 1984. Conservation status of the Malaysian fauna. III. Amphibians. *Malayan Naturalist* 37(4)May: 6-10.

Malaysia has a favorable climate for amphibians but current rate of rain forest destructions puts 70% at risk over time. A lengthy table listing endangered and threatened amphibia by scientific name, English name, Malaysian name, and conservation status is included.

* Kiew, B.H. 1985. Malaysia's 10 most endangered animals, plants and areas. *Malayan Naturalist*. 38(4)May: 2-3.

Cited in **General** Chapter item 51.

848. Kofron, C.P., and A.A. Schreiber. 1985. Ecology of two endangered aquatic

turtles in Missouri: *Kinosternon flavescens* and *Emydoidea blandingii*. *Journal of Herpetology* 19(1): 27-40.

849. Kushlan, J.A. 1988. Conservation and management of the American crocodile. *Environmental Management*. 12(6): 777-790.

Report of a collaborative research effort to determine the causes limiting population growth.

850. Lund, P.F. 1985. Hawksbill turtle (*Eretmochelys imbricata*) nesting on the east coast of Florida. *Journal of Herpetology*. 19(1): 164-166.

851. Luxmoore, R. 1988. The caiman lizard. *Traffic Bulletin*. 9(4): 80-81.

Not seen.

852. Machado, A. 1985. Hypothesis on the reasons for the decline of the large lizards in the Canary Islands. *Bonner Zoological Beitrag*. 36(3-4): 563-575.

Not seen. Article written in English.

853. ______, L.F. Lopez-Jurado, and A. Martin. 1985. Conservation status of reptiles in the Canary Islands, Spain. *Bonner Zoological Beitrag*. 36(3-4): 585-606.

First herpetologia canariensis symposium, Bonn, W. Germany, Nov. 1-3, 1984.

854. Mager, A., Jr. 1985. *Five-year status reviews of sea turtles listed under the Endangered Species Act of 1973*. St. Petersburg, Fla.: U.S. Department of Commerce, National Oceanic and Atmospheric Administration, National Marine Fisheries Service. Government Document No. C 55.302-T 86.

* Majumdar,S.K, F.J. Brenner, and A.F. Rhodes, eds. 1986. *Endangered and threatened species programs in Pennsylvania and other states: causes, issues and management*. Harrisburg, Pennsylvania: Pennsylvania Academy of Science.

Cited in **General** Chapter item 54.

855. Martinez Rica, J.P., and A.M. Cirer Costa. 1982. Notes on some endangered species of Spanish herpetofauna. 1. *Podarcis pityusensis bosca. Biological Conservation* 22(4): 295-314.

856. Mazzotti, F.J. 1984. The ecology of *Crocodylus acutus* in Florida - A thesis in ecology. *Dissertation Abstracts International*. (B)45(1):54.

857. McWilliams, S.R., and M.C. Bachmann. 1988. Using life history and ecology as tools to manage a threatened salamander species. *Journal of the Iowa Academy of Science*. 95(2): 66-71.

858. Metzger, S., and R.W. Marlow. 1986. The status of the Pinzon Island giant tortoise. *Noticias de Galapagos*. No. 43:18-20.

A difficult journal to find.

859. Meylan, A.B. 1983. Marine turtles of the Leeward Islands, Lesser Antiles. *Atoll Research Bulletin.* No. 278, pages 1-43.

After a review of the literature, nesting locations on each island are described noting conservation issues at each site.

860. ______. 1986. Riddle of the ridleys. *Natural History* 95(11)Nov: 90-96.

The migration and reproduction patterns of this most endangered of all sea turtles are examined.

* Mittermeier, R.A. 1981. Brunei protects its wildlife. *Oryx* 16(1)May: 67-70.

Cited in **General** Chapter item 57.

861. Morafka, D.J. 1982. The status and distribution of the bolson tortoise (*Gopherus flavomarginatus). Wildlife Research Report No. 12*. U.S. Fish and Wildlife Service, pp. 71-94.

862. ______. 1988. Ecogeography of the Mexican bolson tortoise (*Gopherus flavomarginatus*): derivation of its endangered status and recommendations for its conservation. Part 3. Historical biogeography of the bolson tortoise. *Annals of the Carnegie Museum* 57(1): 47-72.

863. Mortimer, J.A. 1984. *Marine turtles in the Republic of the Seychelles; status and management: a report on project 1809 (1981-1984).* Gland, Switzerland: IUCN/ World Wildlife Fund.

Bibliography, p. 32-35.

864. Mrosovsky, N. 1983. *Conserving sea turtles.* London: British Herpetological Society.

Extensive bibliography, p. 153-169.

* Mukherjee, A.K. 1982. *Endangered animals of India.* Calcutta: Zoological Survey of India.

Cited in **General** Chapter, item 59.

* National Wildlife Federation. 1988. Twentieth environmental quality index. *National Wildlife* 26(2)Feb-Mar: 38-46.

Cited in **General** Chapter item 62.

865. Nature Conservancy Council (England). 1986. *The conservation of endangered amphibians and reptiles.* [London]: Northminster House.

Not seen.

866. Nawaz, M., and Y. Nawaz. 1986. Status, distribution and conservation of the Afghan Tortoise (*Testudo horsfieldii* Gray, 1844) in Pakistan. *Tigerpa-*

per 13(4): 20-21.

* Nilsson, G. With contributions by M. Gean, J. Buckley, and B. Greveman. 1983. *The endangered species handbook.* Washington, D.C.: Animal Welfare Institute.

Cited in **General** Chapter item 63.

* Norden, A.W., D.C. Forester, and G. Fenwick. 1984. *Threatened and endangered plants and animals of Maryland.* Special Publication/ Maryland Natural Heritage Program: 84-I. Annapolis, Md.: Department of Natural Resources.

Cited in **General** Chapter item 64.

867. Norman, D.R. 1987. Man and Tegu Lizards in eastern Paraguay. *Biological Conservation* 41(1): 39-56.

* Norris, R. 1988. Data for diversity. *Nature Conservancy Magazine.* 38(1)Jan-Feb: 4-10.

Cited in **General** Chapter item 65.

868. Oliver, W.L.R. 1982. The coney and the yellow snake: the distribution and status of the Jamaican Hutia *Geocapromys brownii* and the Jamaican boa *Epicrates subflavus. Dodo* 19:6-33.

869. Orr, L.P. 1984. *A survey of the amphibians and reptiles in the Cuyahoga Valley national recreation area Ohio USA.* 93rd annual meeting of the Ohio Academy of Science on Industry/Academic Relations, Cleveland, Ohio, April 27, 1984. *Ohio Journal of Science.* 84(2): 1-86.

870. Pappas, N.G. 1983. Sea turtles. *Sierra.* 68(SeptOct): 55-58.

Well illustrated, popular level article.

871. Pfannmuller, L. 1986. Status of amphibians and reptiles in Minnesota: legal status. *Occasional Paper, James Ford Bell Museum of Natural History, No. 18.* Minneapolis: University of Minnesota.

* Pilleri, G., and O. Pilleri. 1981. Some observations on mammals, birds and reptiles in the lower valley of the Indus and upper valley of the Brahmaputra. *Bollettino Soc. Adriat. Sci. Nat.* 65(2): 133-166.

Cited in **Mammals** Chapter item 317.

872. Pooley, A.C. 1981. Disappearing African crocodiles. *Oryx* 16(1)May: 38-40.

Reporting countries describe the current status of their crocodile population using terms "endangered, very rare, scarce, seriously depleted, and uncommon." Decline due to killing for many reasons: for hides, by fishermen who see crocodiles as dangerous to fish supply, for food, for export to zoos, as predators, for expansion of agriculture and loss of habitat, more availability of arms and ammunition.

873. Potter, F.E., Jr. 1981. *Status of the American alligator in Texas: special report.* Austin, Texas: Wildlife Division, Texas Parks and Wildlife Dept.

* Quammen, D. 1988. *The flight of the iguana.* New York: Delacorte Press.

Story of extinct and nearly gone lizards on the Galapagos Islands. See also item 680 in the **Birds** Chapter.

874. Raxworthy, C.J. 1988. Reptiles, rain forests and conservation in Madagascar. *Biological Conservation* 43(3): 181-212.

875. Reyes Osorio, S., and R.B. Bury. 1982. Ecology and status of the desert tortoise (*Gopherus agassizii*) on Tiburan Island, Sonora. *Wildlife Research Report No. 12.* U.S. Fish and Wildlife Service. pp. 39-49.

876. Reynolds, R.P., and R.W. Marlow. 1983. Lonesome George, the Pinta Island Tortoise: A case of limited alternatives. *Noticias de Galapagos.* 37:14-17.

* Roots, C. 1987. *Endangered species: Canada's disappearing wildlife.* Illus. C. Godkin. Markham, Ont: Fitzhenry & Whiteside.

Cited in **General** Chapter as item 75.

877. Rose, F.L., and F.W. Judd. 1982. Biology and status of Berlandier's tortoise (*Gopherus berlandieri*). *Wildlife Research Reports No. 12.* U.S. Fish and Wildlife Servive, pp. 57-70.

This entire issue concerns herpetofauna.

* Saharia, V.B. 1982. *Wildlife in India.* Dehra Dun, India: Natraj Publishers.

Cited in **General** Chapter item 76.

878. Samour, H.J. et al. 1987. Survey of the Aldabra giant tortoise population introduced on Curieuse Island, Seychelles. *Biological Conservation.* 41(2): 147-158.

* Sarkar, S.U. 1985. Endangered wildlife of the coastal zone of Bangladesh. *Coastal Zone.* 1985(1):1110-1120.

Cited in **General** Chapter as item 77.

879. Schwab, D. 1982. A list of amphibians and reptiles of the Great Swamp of New Jersey with special attention to state threatened and endangered species. *Bulletin of the Chicago Herpetological Society.* 17(3): 69-76.

880. Senecal Kearney, C., ed. 1981. Pennsylvania species of special concern. *Endangered Species Technical Bulletin.* 6(11): 4-7.

* Skoog, P.J. 1982. *Highways and endangered wildlife in Florida: impacts and recommendations*. (Final report) Tallahassee: Florida Game and Fresh-water Fish Commission.

Cited in **General** Chapter item 82.

881. Stanner, M. 1985. Monitoring the desert monitor. *Israel Land and Nature*. 10(4): 142-145.

* Steadman, D.W., and S. Zousmer. 1988. *Galapagos: discovery on Darwin's islands*. Washington, D.C.: Smithsonian Institution Press.

Cited in **General** Chapter item 85.

* Stermer, D. 1981. *Vanishing creatures, a series of portraits*. Berkeley, Calif.: Lancaster-Miller Publishers.

Cited in **General** Chapter item 86.

882. Sternberg, J., comp. 1982. *Directory of sea turtle conservation programs in the southeastern U.S.* Washington, D.C.: Sea Turtle Rescue Fund/Center for Environmental Education.

A book for reference to current conservation activity. Natural history information is not included.

* Stuart, S.N., ed. 1986. *Conservation of Cameroon montane forests*. Report of the ICBP Cameroon Montane Forest Survey, Nov. 1983-April 1984. Cambridge, England: ICBP.

Information about reptile and amphibian forest species is included. Cited in **General** Chapter item 88.

883. Stubbs, D. 1986. Action programme for protecting *Testudo hermanni* in southern France. *Environmental Conservation* 13(2):167.

884. Sunquist, F. 1988. The private passion of Tomas Blohm. *International Wildlife* 18(1)Jan-Feb: 20-24.

The subject is the alligator.

885. ______. 1988. Zeroing in on the keystone species. *International Wildlife* 18(3)Sept-Oct: 18,20,23.

Florida lists the gopher tortoise (also called the gopher turtle) as "a species of special concern" because of the ecological impact on other species.

* Tikader, B.K. 1983. With a foreword by T.N. Koshoo. *Threatened animals of India*. Calcutta: Zoological Survey of India.

Crocodile is discussed. Cited in **General** Chapter as item 89.

886. Towns, D.R. 1985. The status and prospects of the rare New Zealand lizards *Leiolopisma grande* (gray), *Cyclodina whitakeri hardy* and *Leiolopisma*

otagense mccann (Lacertilia: Scincidae). In *Biology of Australian frogs and reptiles,* ed. G. Grigg, R. Shine, and H. Ehmann, 481-489. Chipping Norton (NSW): Surrey Beatty & Sons and the Royal Zoological Society of New South Wales.

887. Turner, J.L. 1986. A tale of sea turtles. *Conservationist.*(New York) 41(2)Sept-Oct: 9-13, 56.

Describes the problem of strandings on Long Island beaches.

* U.S. Congress. House. Committee on Merchant Marine and Fisheries. Subcommittee on Fisheries and Wildlife Conservation and the Environment. 1987. *Endangered Species Act reauthorization: Hearings on H.R. 1467.* 100th Cong., 1st sess., 1987. (March 17). Washington, D.C.: Superintendent of Documents, G.P.O.

This 725 page volume includes statements by legislators on the committee, and exhibits and statements of an educational nature by interested organizations and individuals intended to persuade passage. Marine turtles were prominent among the animals of concern. The exhibits may be articles and excerpts from books and contain much information. Cited as item 94.

888. Uchida, I. 1986. Disruption of loggerhead sea turtle nesting crawls with emphasis on the barrier in the sea. Marine Chelonological Notes 3. *Aquabiology* 8(3): 217-219.

889. ______. 1983. Manual for the marine chelonology 3. Growth. *Aquabiology.* 5(1): 36-42.

890. Van Rhijn, F.A., and J.C. Van Gorkom. 1983. Optic orientation in hatchlings of the sea turtle, *Chelonia mydas.* 3. Sea finding behavior: the role of photic and visual orientation in animals walking on the spot under laboratory conditions. *Marine Behavior Physiology* 9(3): 211-228.

The sea turtle is at its most vulnerable during the period just after hatching which is the focus of this article.

* Vandenbeld, J. 1988. *Nature of Australia; a portrait of the island continent.* Published in association with the Australian Broadcasting Association. New York: Facts on File.

Comprehensive view of Australian nature extending to New Guinea. See **General** Chapter item 99.

891. Varona, L.S. 1981. Cuban crocodiles. *Rephiberary.* 40:7-8.

Not seen. A difficult journal to find.

* Wallace, D.R. 1987. *Life in the balance.* New York: Harcourt Brace Jovanovich.

Popular, illustrated account noting decline of sea turtles. See **General** Chapter item 101.

892. Watanabe, M.E. 1986. A change of fortune for the Chinese alligator. *Animal Kingdom* 89(5)Sept/Oct: 34-39.

The smallish Chinese alligator is unique to China and the rarest of all crocodiles. Its range has shrunk 90%. Possible to breed in captivity but may need protected undisturbed habitat. Article relates reproduction and nesting habits. Giant saltwater crocodile probably extinct in China having not been seen in 50 years.

893. Webb, G.J.W., S.C. Manolis, and P.J. Whitehead, eds. 1987. *Wildlife management: crocodiles and alligators*. Chipping Norton, NSW, Australia: Surrey Beatty and Sons, Ltd.

Collection of articles describing management patterns worldwide. Includes a list of species considered rare and endangered by IUCN and brief descriptions.

894. Weller, W. 1982. Rare and endangered reptiles in Ontario. Part 1. *Bulletin of Canadian Amphibian and Reptile Conservation Society*. 20(2): 1-8.

Brief species descriptions of 14 reptiles the author considers to be in jeopardy in Ontario. A brief overview of the development of Canada's endangered species list is included, names of organizations that were involved, and definitions of the five levels of status categories for species of concern.

895. Wexler, M. 1988. Modern mission to save an ancient mariner. *National Wildlife*. 26(4): 4-11.

The subject is sea turtles.

* White, C.P. 1982. *Endangered and threatened wildlife of the Chesapeake Bay Region: Delaware, Maryland, and Virginia*. Centerville, Md.: Tidewater Publishers.

Turtles among the many animals discussed. Cited in **General** Chapter item 105.

* Wilson, J.H., Compiler and Editor. 1984. *Rare and endangered species of Missouri*. Illus. C.E. Schwartz/ Design and maps by D. Pasley. [Jefferson City, MO.]: Conservation Commission of the State of Missouri.

Lists 14 amphibians and reptiles. Cited also in **General** Chapter item 108.

896. Wilson, S.K., and D.C. Knowles. 1988. *Australia's reptiles: a photographic reference to the terrestial reptiles of Australia*. Sydney: Collins Press.

Endangerment not always noted.

897. Witzell, W.N. 1985. Variation of size at maturity of female hawksbill turtles (*Eretmochelys imbricata*) with speculations on life-history tactics relative to proper stock management. *Japanese Journal of Herpetology* 11(2):46-51.

898. Wood, V.E. 1986. Breeding success of hawksbill turtles *Eretmochelys*

imbricata at Cousin Island, Seychelles and the implication of their conservation. *Biological Conservation* 37(4): 321-332.

* Woolfenden, G.E. 1983. *Rare, threatened, and endangered vertebrates of southwest Florida and potential OCS activity impacts.* Washington, D.C.: U.S. Fish and Wildlife Service, Office of Biological Services, and Bureau of Land Management, Branch of Offshore Studies.

Cited in **General** Chapter item 112.

* Yaffee, S.L. 1982. *Prohibitive policy: implementing the federal Endangered Species Act.* MIT Studies in American politics and public policy, 9. Cambridge, MA: MIT Press

Green turtles, loggerheads and kemp ridleys are the reptiles described. Cited in **General** Chapter item 113.

898a. Zhujian, H. 1982. The Chinese salamander. *Oryx* 16(3)Feb: 272-273.

A protected species in China, this salamander is the largest now in existence and has a wide range of distribution. Declining numbers have caused government action.

899. Zug, G.R., A.H. Wynn, and C. Ruckdeschel. 1986. *Age determination of loggerhead sea turtles, Caretta caretta, by incremental growth marks in the skeleton.* Smithsonian Contributions to Zoology, n. 427. Washington, D.C.: Smithsonian Institution Press.

FISHES

900. Allendorf, F.W., and R.F. Leary. 1988. Conservation and distribution of genetic variation in a polytypic species, the cutthroat trout. *Conservation Biology* 2(2)June: 145-148.

Eleven subspecies have protected legal status in one or more states. Paper points to need for maintaining genetic diversity of this native species and its subspecies. A major danger to American trout in the west has been the introduction of nonnative species. The resulting interbreeding causes "hybrid swarms" which the author opposes. Paper is a scientific analysis of studying divergencies within taxa and between taxa.

901. Balon, E.K., M.N. Bruton, and H. Fricke. 1988. A fiftieth anniversary reflection on the living coelacanth, *Latimeria-chalumnae*. Some new interpretations of its natural history and conservation status. *Environmental Biology of Fishes* 23(4): 241-280.

* Berger, T.J., A.M. Neuner, and S.R. Edwards. 1981. *Directory of federally controlled species*. Lawrence,Kansas: Association of Systematics Collections.

Cited in **General** Chapter as item 11.

902. Boschung, H. 1986. Pygmy sculpin. In *Vertebrate animals of Alabama in need of special attention*. ed. R.H. Mount, 11-12. Auburn Alabama: Alabama Agricultural Experiment Station.

The pygmy sculpin is described.

* Boyd, J.M. 1984. Nature conservation. *Proceedings of the Royal Society of Edinburgh*. Section B-Biological Sciences. 84: 295-335.

See **General** Chapter, item 12.

* Campbell, R.R. 1984. Rare and endangered fishes of Canada: The Committee on the Status of Endangered Wildlife in Canada (COSEWIC) Fish and Marine Mammals Subcommittee. *Canadian Field-Naturalist*. 98(1)Jan/Mar: 71-74.

Describes the function of COSEWIC (the Canadian "endangered species list") and provides a list of fishes and marine mammals that are on the list. A table listing those species under study makes this article a useful starting point for identifying endangered fish and marine species. Also cited in **Mammals** Chapter as item 148.

* Campbell, R.R. 1985. Rare and endangered fishes and marine mammals of Canada: COSEWIC Fish and Marine Mammals Subcommittee status report: 2. *Canadian Field-Naturalist.* 99(3):404-408.

Follow-up to a series of status reports published as articles in the *Canadian Field-Naturalist,* 98(1) 1984, that analyzed the population of candidates for listing on COSEWIC, the Canadian endangered species list. Author also introduces seven status reports of individual species. Article includes tables of fish and marine mammal species that have been assigned COSEWIC status and those that are of interest to COSEWIC.

Cited in **Mammals** as item 149.

* Campbell, R.R. 1987. Rare and endangered fishes and marine mammals of Canada: COSEWIC Fish and Marine Mammal Subcommittee status reports, 3. *Canadian Field-Naturalist.* 101(2): 165-170.

Cited in **Mammals** Chapter as item 150.

* Chandler, W.J., and L. Labate, Eds. 1988. *Audubon wildlife report 1988/1989.* San Diego: Academic Press.

This issue's special report examines the U.S. National Marine Fisheries. Cited in **General** Chapter as item 16.

* Coffin, B., and L. Pfannmuller, eds. 1988. *Minnesota's endangered flora and fauna.* Illus. by Jan A. Janssens. Minneapolis: Univ. of Minnesota Press for the Natural Heritage and Nongame Wildlife Programs of the Division of Fish and Wildlife, Minnesota Department of Natural Resources.

Cited in **General** Chapter item 17.

903. Collins, J.P., C. Young, J. Howell, and W.L. Minckley. 1981. Impact of flooding in a Sonoran desert stream, including elimination of an endangered fish population. *Southwestern Naturalist* 26(4) Nov. 20: 415-423. Austin, Texas.: Southwestern Assn. of Naturalists.

The Gila Topminnow is described.

* Convention on International Trade in Endangered Species (CITES). 1981- *Identification manual.* Ed., P. Dollinger. 5 v. (loose-leaf) v.3. Reptilia. Amphibia. Pisces. Gland, Switzerland: Secretariat of the Convention.

See item 19 in **General** Chapter.

* Cortese, T.J., and B.A. Grosnek, Compilers. 1987. *Selected publication series of the U.S. Fish and Wildlife Service, 1889-1985.* U.S. Fish and Wildlife Service, Resource Publication 159. Washington, D.C.: U.S. Department of the Interior, Fish and Wildlife Service.

See **General** Chapter item 20.

* DiSilvestro, R.L., and A.S. Eno, eds. 1986. *Audubon wildlife report 1986.* New York: National Audubon Society. (Dist. by Academic Press).

The Atlantic and chinook salmon are described. See item 26 in **General** Chapter.

* DiSilvestro, R.L., ed. 1987. *Audubon Wildlife Report 1987*. Orlando: National Audubon Society (Dist. by Academic Press)

The paiute trout is one of the fishes discussed.

See item 27 in **General** Chapter.

* Doughty, R.W. 1983. *Wildlife and man in Texas*. College Station: Texas A & M University Press.

Cited in **Mammals** Chapter item 180.

904. Evans, W.A. 1981. Can we save our anadromous fish resources in small coastal streams? -- a look at the past, present, and future. *Cal-Nev Wildlife*. 1981: 148-154.

* Genoways, H., and F.J. Brenner, eds. 1985. *Species of special concern in Pennsylvania*. Compiled by the Pennsylvania Biological Survey. Special Publication of the Carnegie Museum of Natural History, No. 11. Pittsburgh: Carnegie Museum of Natural History.

A chapter on fish is included. See item 35 in **General** Chapter.

* Groves, R.H., and W.D.L. Ride, eds. 1982. *Species at risk: research in Australia*. Proceedings of a Symposium on the Biology of Rare and Endangered Species in Australia held in Canberra, November 25 and 26, 1981.

See **General** Chapter item 37.

905. Hamman, R.L. 1982. Spawning and culture of bonytail (1) and humpback (2) chub. *Progressive Fish-Culturist* 44(4): 201-203; 213-216.

Increasing the population of an endangered fish through artificial culture.

906. Harris, J.H. 1986. Rare and endangered: classifying Australia's threatened fishes. *Australian Natural History*. 22(5)Winter: 208-209.

Trout cod, Eastern freshwater cod and others.

907. Hubbs, C., and B.L. Jensen. 1984. Extinction of *Gambusia amistadensis*, an endangered fish. *Copeia*. 1984(2): 529-530.

* International Convention Advisory Commission. 1981. *Convention on international trade in endangered species of wild fauna and flora*. Washington, DC:

See item 44 in **General** Chapter.

* IUCN Conservation monitoring centre. 1986. *1986 IUCN red list of threatened animals*. Prepared by the IUCN Conservation Monitoring Centre, Cambridge, U.K.

See item 44 in **General** Chapter.

* IUCN Conservation Monitoring Centre. 1988. *1988 IUCN red list of threatened animals.* Cambridge, U.K., and Gland, Switzerland: IUCN.

See item 45 in **General** Chapter.

908. Johnson, J.E. with editorial support from B.D. McAleer. 1987. *Protected fishes of the United States and Canada.* Bethesda: American Fisheries Society.

Useful for identification. Lists of protected species by family include common name and genera species name. Protected fishes within each state and province are listed by popular name.

909. Kellert, S.R. 1986. Social and perceptual factors in the preservation of animal species. In *Preservation of species: the value of biological diversity.* ed., B.G. Norton, pp. 50-73. Princeton: Princeton Univ. Press.

Snail darter used as a case study to address philosopical and management aspects of preservation.

* Kiew, B.H. 1985. Malaysia's 10 most endangered animals, plants and areas. *Malayan Naturalist.* 38(4)May: 2-3.

The kelesa (Golden Dragon Fish) is considered. See **General** Chapter item 51.

910. Lelek, A. 1987. *Threatened fishes of Europe.* The Freshwater Fishes of Europe. Vol.9. Wiesbaden: Aula-Verlag.

Identifies fish considered endangered or threatened in some countries or rivers but may have healthy populations in other regions.

911. Lindquist, D.C., and L.M. Page, eds. 1984. *Environmental biology of darters.* Papers from a symposium on the comparative behavior, ecology, and life histories of darters (Etheostomatini) held during the 62nd annual meeting of the American Society of Ichthyologists and Herpetologists at DeKalb, Illinois, June 14-15, 1982. Reprinted in *Environmental Biology of Fishes* with four new papers. 11(2): 83-159. The Hague: Junk.

* Majumdar, S.K, F.J. Brenner, and A.F. Rhodes, eds. 1986. *Endangered and threatened species programs in Pennsylvania and other states: causes, issues and management.* Harrisburg, Pennsylvania: Pennsylvania Academy of Science.

See item 54 in **General** Chapter.

912. Mayer, K.S., E.P. Multer, and R. Kent. 1984. *Acid rain: effects on fish and wildlife.* Washington, D.C.: U.S. Fish and Wildlife Service.

913. McAllister, D.E., B.J. Parker, and P.M. McKee. 1985. Rare, endangered and extinct fishes in Canada. *Syllogeus.* No. 54. Ottawa: National Museums of

Natural Sciences, National Museums of Canada.

Summary in French. Lenthy bibliography.

914. Meffe, G.K., and R.C. Vrijenhoeck. 1988. Conservation genetics in the management of desert fishes. *Conservation Biology*. 2(2)June: 157-169.

Aruges for need for understanding genetic population structure as river dwelling desert fishes frequently exist in colonies. Concerned that habitat distruction will destroy the linkage sites between the colonies causing the colonies to become isolated and threatening the necessary genetic diversity. Describes the endangerment of the Sonoran top minnow through the introduction of the mosquito fish.

915. Miller, W.H., H.M. Tyus, and C.A. Carlson, eds. 1982. *Fishes of the Upper Colorado River system, present and future*. Proceedings of a symposium presented at the annual meeting of the American Fisheries Society in Albuquerque, New Mexico, September 18, 1981. Bethesda, Md.: American Fisheries Society, Western Division.

Pressures on the Upper Colorado River system are a cause of concern for endemic fish survival.

916. Moyle, P.B., and F.R. Senanayake. 1981. Conservation of ornamental freshwater fishes in Sri Lanka. *Tigerpaper*. 8(4): 1-3.

The export of ornamental fishes is a significant part of the economy of Sri Lanka. Their abundance is an indicator of conditions of the rivers. Lists 15 endemic species and factors affecting their status.

* Nilsson, G. With contributions by M. Gean, J. Buckley, and B. Greveman. 1983. *The endangered species handbook*. Washington, D.C.: Animal Welfare Institute.

Cited in **General** Chapter as item 63.

917. Noakes, E.L.G., ed. 1983. Predators and prey in fishes. Proceedings of the 3rd biennial conference on the Ethology and Behavioral Ecology of Fishes held at Normal, Illinois, May 19-22, 1981. *Environmental Biology of Fishes*. 8(3-4).

* Norden, A.W., D.C. Forester, and G.H. Fenwick. 1984. *Threatened and endangered plants and animals of Maryland*. Annapolis, Md.: Department of Natural Resources.

See item 64 in **General** Chapter.

918. Ono, R.D., J.D. Williams, and A. Wagner. 1983. *Vanishing fishes of North America*. Paintings by Aleta Pahl. Washington, D.C.: Stone Wall Press. Distributed by Stackpole Books, Harrisburg, Pa.

An illustrated book covering nearly 60 extinct, endangered, and threatened fishes. Explores causes of extinction.

919. Pearce, W.A. 1986. Methuselah of freshwater fishes: the lake sturgeon. *Conservationist* (New York) 41(3)Nov-Dec: 11-13.

Almost decimated in the 1880's, fish is now protected in New York as a threatened species. Article describes biology, range, size and long life.

* Peterson, M.A., and D. Adler. 1982. *The effects of air pollution and acid rain on fish, wildlife, and their habitats: critical habitats of threatened and endangered species*. U.S. Fish and Wildlife Service, Office of Biological Service, Report No. 11. (Air pollution and acid rain; Report No. 11). (The effects of air pollution and acid rain on fish, wildlife, and their habitats; v.9). Washington, D.C.: GPO.

See also **General** Chapter item 70.

920. Pister, E.P., and J.H. Kerbavaz. 1984. Fish slough: a case study in management of a desert wetland system. In *California riparian systems: ecology, conservation and productive management*. eds. R.E. Warner and K.M. Hendrix, 929-933. Berkeley, Los Angeles, and London: Univ. of California Press.

Habitats as breeding areas are described.

921. Ramsey, J.S. 1986. Paleband shiner. *Vertebrate animals of Alabama in need of special attention*. ed., R.H. Mount, 6-7. Auburn, Alabama: Alabama Agricultural Experiment Station.

922. ______, and H. Boschung. 1986. Slackwater darter. *Vertebrate animals of Alabama in need of special attention*. ed., R.H. Mount, 7-8. Auburn, Alabama: Alabama Agricultural Experiment Station.

923. ______, and J.M. Pierson. 1986. Blue shiner. *Vertebrate animals of Alabama in need of special attention*. ed., R.H. Mount, 12-13. Auburn, Alabama: Alabama Agricultural Experiment Station.

924. Reiger, G. 1982. Fishing: blue tuna in dire straits. *Audubon* 84(3)May: 112-117.

925. Rinne, J.N. 1984. *Endangered and nongame fish programs in the west: their role in the conservation of native fishes*. Proceedings, 64th Annual Conference of the Western Association of Fish and Wildlife Agencies. 64:456-469.

* Roots, C. 1987. *Endangered species: Canada's disappearing wildlife*. Illus. C. Godkin. Markham, Ont: Fitzhenry & Whiteside.

Twenty-two species of fish are described. See **General** Chapter, item 75.

926. Schmidt, K.P. 1984. *Specially protected fishes*. American Currents: a publication of the North American Native Fishes Association, Special Edition. Marine Education Center, 1650 E. Beach Drive, Biloxi, Miss. 39530.

A mimeographed publication listing the state and Canadian provincial lists of endangered, threatened and special-status species. The Federal list is not included. Contents are common and scientific names only.

927. Scoppettone, G., M. Coleman, and G.A. Wedemeyer. 1986. *Life history and status of the endangered cui-ui of Pyramid Lake, Nevada.* Fish and Wildlife Research 1. Fort Collins, CO: U.S. Fish and Wildlife Service.

A 23-page pamphlet that is specific for this fish in this lake.

928. Sheldon, A. 1988. Conservation of stream fishes: patterns of diversity, rarity, and risk. *Conservation Biology* 2(2)June: 149-156.

Cites the increase in concern for conservation by the twin approaches of legal protection for single species and habitat preservation but the author argues that river systems and the impact of navigational and hydroelectric projects should be examined for effects on species moving into new drainage areas and competing successfully against endemic species.

* Silber. G.K. 1987. Can the vaquita be saved? *Defenders* 62(3)May-June: 14-15.

Cited in **Mammals** Chapter as item 358. Author also describes a totoaba (also called seatrout or weakfish) fishery which is operating even though this species is also endangered.

929. U.S. Fish and Wildlife Service. 1985. Determination that the Warner sucker is a threatened species and designation of its critical habitat. (Final Rule) *Federal Register,* September 27, 1985. 50(188): 39117-29.

930. U.S. Fish and Wildlife Service. 1985. *Endangered and threatened wildlife and plants; proposal to list the Waccamaw silverside as a threatened species and designation of its critical habitat.* (Proposed Rule) Federal Register November 7, 1985. 50(216): 46320-4.

931. Utter, F.M. 1981. Biological criteria for definition of species and distinct intraspecific populations of anadromous salmonids under the U.S. Endangered Species Act of 1973. *Canadian Journal Aquatic Science* 38(12): 1626-1635.

932. Vrijenhoek, R.C., M.E. Douglas, and G.K. Meffe. 1985. Conservation genetics of endangered fish populations in Arizona. *Science* 229(4711): 400-402.

933. Warren, M.L., Jr., and B.M. Burr. 1988. Reassessment of the Illinois USA ranges of the bigeye chub (*Hybopsis-amblops*) and the pallid shiner (*Notropis-amnis.*) *Ohio Journal of Science* 88(5): 181-183.

* White, C.P. 1982. *Endangered and threatened wildlife of the Chesapeake Bay Region: Delaware, Maryland, and Virginia.* Centerville, Md.: Tidewater Publishers.

The chub, darter, and madtom are the fishes described. See also item 105 in **General** Chapter.

934. Williams, J.D. 1981. Threatened desert fishes and the Endangered Species Act. In *Fishes in North American deserts*. eds. E.P. Pister, R.J. Naiman, and D.L. Soltz, 447-475. New York: John Wiley.

A table of endangered and threatened fishes of the U.S. is accompanied by an excellent summary of the history of the movement that culminated in the Endangered Species Act of 1973. Includes a fine bibliography.

935. Williams, J.E., and J.E. Deacon. 1986. Subspecific identity of the Amargosa pupfish, *Cyprinodon nevadensis,* from Crystal Spring, Ash Meadows, Nevada. *Great Basin Naturalist* 46(2): 220-223.

936. Williams, J.E., and D.W. Sada. 1985. Status of two endangered fishes, *Cyprinodon nevadensis mionectes* and *Rhinichthys osculus nevadensis,* from two springs in Ash Meadows, Nevada. *Southwestern Nature*. 30(4): 475-484.

* Wilson, J.H., Compiler and Editor. 1984. *Rare and endangered species of Missouri*. [Jefferson City, MO.]: Conservation Commission of the State of Missouri.

Thirty-five fish listed. See item 108 in **General** Chapter.

* Woolfenden, G.E. 1983. *Rare, threatened, and endangered vertebrates of southwest Florida and potential OCS* (Outer Continental Shelf) *activity impacts*. FWS/OBS 82/03. Washington, D.C.: U.S. Fish and Wildlife Service, Office of Biological Services, and Bureau of Land Management, Branch of Offshore Studies.

Cited in **General** Chapter as item 112.

* Yaffee, S.L. 1982. *Prohibitive policy: implementing the Federal Endangered Species Act*. MIT Studies in American politics and public policy, 9. Cambridge, MA: MIT Press.

Legal aspects of the snail darter. See **General** Chapter item 113.

937. Yan, Hong-Young. 1986. Reproductive strategies of the clear creek gambusia, *Gambusia heterochir*. *Dissertation Abstracts International* 47/05-B: 1896-2084. Ph.D. 1986, The University of Texas at Austin.

A scientific investigation of this endangered fish examined effects of maturation and body size in males and females on fecundity.

ADDENDA

938. Beiswenger, R.E. 1986. An endangered species, the Wyoming toad, *Bufo hemiophrys baxteri*—the importance of an early warning system. *Biological Conservation* 37(1): 59-71.

939. Clark, B.K., J.B. Bowles, B.S. Clark. 1987. Summer status of the endangered Indiana bat in Iowa. *American Midland Naturalist* 118(1)July: 32-39.

An intensive study and report.

940. Dixon, K.R., and T.C. Joelson. 1987. The political economy of the spotted owl. *Ecology* 68(4)Aug: 772-776.

Recommendations include moratorium on cutting old growth forest while debate continues.

941. Levee, D. 1988. Why is the houbara *Chlamydotis undulata macqueenii* still an endangered species in Israel? *Biological Conservation* 45(1): 47-54.

942. Levine, N.D. 1986. Preservation versus elimination. *Bioscience* 36(5)May: 308-309.

Argues that extinction is part of evolution. Points out contradiction of preserving some species while trying to eradicate others, e.g., roaches, smallpox and polio organisms. No grief for the passing of a snail darter or condor. Article began an exchange of articles and letters that examined a controversial issue.

943. Marx, W. 1988. Vanishing vaquitas. *Oceans* 21(Sept/Oct): 52-55.

944. Newman, A., M. Bush, D.E. Wildt, D. Van Dam, M.T. Frankenhuis, L. Simmons, L. Phillips, and S.J. O'Brien. 1985. Biochemical genetic variation in eight endangered or threatened felis species. *Journal of Mammalogy* 66(2)May: 256-267.

Use of electrophoresis to examine genetic variation at 25 biochemical loci using zoo cats.

945. Pimm, S.L., H.L. Jones, and J. Diamond. 1988. On the risk of extinction. *American Naturalist* 132(6)Dec: 757-785.

A highly theoretical analysis concerning variables in population biology that affect risk using bird censuses as models. Several tables of species included.

946. Saleh, M.A. 1987. The decline of gazelles in Egypt. *Biological Conservation* 39(2): 83-95.

947. Salwasser, H. 1987. Spotted owls turning a battle-ground into a blueprint. *Ecology* 68(4)Aug: 776-779

Describes other species 'victories' as small. Suggests conservation strategy for spotted owl put risks on a scale of importance for society to debate and decide. This last of three commentaries includes a fine bibliography.

948. Simberloff, D. 1987. The spotted owl fracas: mixing academic, applied, and political ecology. *Ecology* 68(4)Aug: 766-772.

Discusses U.S. Forest Service's "Preferred Alternative" of 550 habitat areas.

949. Temple, S.A. 1988. Recovery of the endangered Mauritius kestrel from an extreme population bottleneck. *Auk* 103(July): 632-633.

Shift from nesting in tree cavities to nesting in holes and cliff niches may be the cause of successful breeding.

950. Wilcox, B.A., and D.D. Murphy. 1985. Conservation strategy: the effects of fragmentation on extinction. *American Naturalist* 125:879-887.

A critical comment on a basic issue of population biology: will a Single Large nature reserve Or Several Small (SLOSS) reserves preserve more species. A good bibliography is included.

APPENDIX

Organizations concerned with Endangered Species.

Organization	Serial Publications
American Association of Zoological Parks and Aquariums. Rt 88, Oglebay Park Wheeling, WV 26003	AAZPA MANUAL OF FEDERAL WILDLIFE REGULATIONS. Annual loose-leaf format
American Forestry Assn 18th Street, NW Washington, D.C. 20036	AMERICAN FORESTS
Center for Action on Endangered Species 175 West Main Street Ayer, MA 01432	ENDANGERED SPECIES TECHNICAL BULLETIN
Center for Environmental Education 1925 K Street, NW #206 Washington, D.C. 20006	THE WHALE REPORT ENVIRONMENTAL EDUCATION REPORT
Charles Darwin Foundation for the Galapagos Isles c/o Marsha E. Sitnik National Museum of Natural History 10th & Constitution Ave., NW Washington, D.C. 20566	NOTICIAS DE GALAPAGOS
Defenders of Wildlife 1244 Nineteenth Street, NW Washington, D.C. 20036	DEFENDERS
Ducks Unlimited One Waterfowl Way Long Grove, IL 60047	DUCKS UNLIMITED

Organization	Publications
Ecological Society of America Center for Environmental Studies Arizona State University Tempe, AZ 85287	ECOLOGICAL MONOGRAPHS
European Information Centre for Nature Conservation B.P. 431R6, 67006 Strasbourg Cedex, France	NATUROPA: Bulletin of the European Information Centre for Nature Conservation.
Fauna and Flora Preservation Society Osney Mead, Oxford OX2 OEL, England	ORYX
Friends of Animals 11 West 60th Street New York, N.Y. 10023	FRIENDS OF ANIMALS (Newsletter)
Friends of the Earth Trent University Peterborough, Ontario K9J 7B8, Canada	ALTERNATIVES
Greenpeace 1436 U Street, N.W. P.O. Box 3720 Washington, D.C. 20007	GREENPEACE
IIED-Earthscan 3 Endsleigh Street London WC1H ODD, UK	TROPICAL FOREST POLICY PAPERS
International Council for Bird Preservation (same as IUCN, see below)	TECHNICAL PUBLICATION SERIES RED DATA BOOKS (in cooperation with the IUCN) U.S. BIRDWATCH 1987- *BUSTARD STUDIES: JOURNAL OF THE ICBP BUSTARD GROUP *Working groups for a particular species often publish a series.

International Crane Foundation E-11376 Shady Lane Rd. Baraboo, Wisconsin	ICF BUGLE
International Foundation for the Conservation of Birds 11300 Weddington Street North Hollywood, CA 91601	PROCEEDINGS
International Union for the Conservation of Nature and Natural Resources (IUCN) Conservation Monitoring Ctr. 219 (c) Huntingdon Road Cambridge CB3 ODL, U.K. and Avenue du Mont-Blanc CH-1196 Gland, Switzerland	THE IUCN RED DATA BOOKS NATIONAL CONSERVATION STRATEGIES NATURAL RESOURCES EXPERTISE PROFILES SPECIES SURVIVAL COMMISSION PUBLICATIONS THE IUCN BULLETIN TRAFFIC BULLETIN PARKS, PARQUES, PARCS SPECIES Newsletter of the Species Survival Commission PRIMATE CONSERVATION
Izaak Walton League of America 1401 Wilson Blvd., Level B Arlington, VA 22209	OUTDOOR AMERICA
Jersey Wildlife Preservation Trust Les Augres Manor Trinity, Jersey Channel Islands, England	DODO DISPATCH DODO JOURNAL (Published in affiliation with the Wildlife Preservation Trust International)
National Audubon Society 950 Third Avenue New York, N.Y. 10022	AUDUBON AUDUBON WILDLIFE REPORTS AMERICAN BIRDS
National Parks and Conservation 1015 31st Street, NW Washington, D.C. 20007	NATIONAL PARKS

National Wildlife Federation 1412 16th Street Vienna, VA 22180	NATIONAL WILDLIFE INTERNATIONAL WILDLIFE CONSERVATION DIRECTORY CONSERVATION REPORT
Natural Resources Defense Council (NRDC) 122 East 42nd Street New York, N.Y. 10017	THE AMICUS JOURNAL NRDC NEWS NRDC DOCKET NRDC Newsletter NRDC CITIZENS ACTION PROJECT BULLETIN
The Nature Conservancy 1800 North Kent Street Arlington, VA 22209	NATURE CONSERVANCY NEWS
North American Loon Fund R.R. 4, Box 240C Meredith, NH 03253	THE LOON CALL
Royal Forest and Bird Protection Society of New Zealand Box 631 Wellington, New Zealand	FOREST AND BIRD
Scientists' Institute for Public Information 4000 Albemarle Street, NW Washington, D.C. 20016	ENVIRONMENT
The Sierra Club 530 Bush Street San Francisco, CA 94109	SIERRA SIERRA CLUB BULLETIN
Smithsonian Institution Department of Botany NHB 166 Washington, D.C. 20560	BIOLOGICAL CONSERVATION NEWSLETTER
Office of Endangered Species U.S. Fish & Wildlife Service Department of the Interior 18th and C Streets, NW Washington, D.C. 20240	LIAISON CONSERVATION DIRECTORY FOR ENDANGERED AND THREATENED SPECIES. 1983, 5th ed. ENDANGERED SPECIES TECHNICAL BULLETIN* (Distributed to government agencies and major GPO repository libraries)

School of Natural Resources University of Michigan Ann Arbor, MI 48109-1115	ENDANGERED SPECIES UPDATE includes *ENDANGERED SPECIES TECHNICAL BULLETIN REPRINT (Reprint of the FWS publication made available for public subscription)
The Wilderness Society 1901 Pennsylvania Avenue, NW Washington, D.C. 20006	WILDERNESS MAGAZINE
Wildlife Conservation International New York Zoological Society The Zoological Park Bronx, N.Y. 10460	ANIMAL KINGDOM
Wildlife Conservation Society of Zambia P.O. Box 30255 Lusaka, Zambia	BLACK LECHWE
Wildlife Management Institute 1101 14th Street, N.W. Suite 725 Washington, D.C. 20005	WILDLIFE TRANSACTIONS OF NORTH AMERICAN WILDLIFE AND NATURAL RESOURCES (Annual)
Wildlife Preservation Trust International Zoological Society of Philadelphia 34th Street and Girard Ave. Philadelphia, PA 19104	DODO DISPATCH (Newsletter) DODO (See above "Jersey Wildlife Preservation Trust")
The Wildlife Society 5410 Grosvenor Lane Bethesda, MD 20014	WILDLIFE MONOGRAPHS JOURNAL OF WILDLIFE MANAGEMENT WILDLIFE SOCIETY BULLETIN
Wolf Sanctuary P.O. Box 760 Eureka, MO 63025	BULLETIN (Qrtly)
World Pheasant Association P.O. Box 5 Lower Basildon, Reading Berkshire RG8 9PF, England	JOURNAL OF THE WORLD PHEASANT ASSOCIATION PROCEEDINGS (Irregular) WPA NEWS

World Pheasant Association of the U.S.A. c/o Eugene Knoder 752 Swede Gulch Road Golden, CO 80401	
World Wildlife Fund 1250 24th St., NW Washington, D.C. 20037	FOCUS WORLD WILDLIFE REPORTS (Irregular) PRIMATE CONSERVATION JOURNAL
Worldwatch Institute 1776 Massachusetts Ave., NW Washington, D.C. 20036	STATE OF THE WORLD WORLDWATCH WORLDWATCH PAPERS

INDEX TO COMMON NAMES

INDEX TO SCIENTIFIC NAMES

AUTHOR INDEX

GEOGRAPHIC AREA INDEX